FOUNDATION CORE 1–3 ANSWERS

Page 1   Task M1.1/M1.2

1. a 20   **b** 500   **c** 100   **d** 4000   **e** 4

2. a 345   **b** 543

3. £50   **4.** £700

5. a $\frac{9}{100}$ **b** $\frac{3}{100}$ **c** $\frac{8}{1000}$ **d** $\frac{5}{10}$

    e $\frac{4}{1000}$ **f** $\frac{6}{10}$ **g** 20   **h** $\frac{7}{1000}$

6. 0·03   **7.** $40 + 9 + \frac{3}{10} + \frac{7}{100}$

8. 0·6   **9.** 0·008   **10.** 48·09

Page 1   Task M1.3/M1.4

1. a 40   **b** 40   **c** 80   **d** 80   **e** 330

2. a 300   **b** 800   **c** 400   **d** 1200   **e** 1600

3. a 4000   **b** 6000   **c** 7000   **d** 4000   **e** 7000

4. £14·00   **5.** 54 kg

6. a 7   **b** 5   **c** 8   **d** 13   **e** 19

7. a 28   **b** 30   **c** 16   **d** 36

    e 110   **f** 14   **g** 0   **h** 12

8. 260, 349, 302, 287

9. a 16 460   **b** 16 500   **c** 16 000

10. £3150

Page 2   Task M1.5/M1.6

1. 111   **2.** 803   **3.** 7513   **4.** 27   **5.** 35

6. 217   **7.** 279   **8.** 3242   **9.** 1811   **10.** 51 380

11. 906   **12.** 281   **13.** 726   **14.** £279   **15.** £739

16. 416   **17.** 427   **18.** 13   **19.** 229   **20.** 170

21. 118   **22.** 204   **23.** 385   **24.** 747   **25.** 8

Page 3   Task M1.7

1. a 7200   **b** 41 600   **c** 58 600   **d** 6730

    e 57 000   **f** 6 720 000

2. a 4700   **b** 38   **c** 480   **d** 100

    e 10   **f** 720   **g** 215, 2150, 215 000

3. £80 000

4. a 1200   **b** 27 000

5. £2 700 000

6. a 80   **b** 40   **c** 6   **d** 300   **e** 800   **f** 21 000

7. £1 240 000, £240 000 more

8. 600, ÷20, 30, ×40, 1200, ×20, 24000, ÷60, 400

Page 4   Task M1.8

1. 168   **2.** 315   **3.** 296   **4.** 147

5. 504   **6.** 912   **7.** 2104   **8.** 3704

9. 4428   **10.** 2556   **11.** £3774   **12.** £120

13. 1904   **14.** £864   **15.** $38 \times 6 \times 8$ is larger by 60

Page 4   Task M1.9

1. 448   **2.** 408   **3.** 966   **4.** 2176   **5.** 3195

6. 15 156   **7.** 23 492   **8.** 35 112   **9.** £592   **10.** 624

11. £4076   **12.** £413

Page 5   Task M1.10

1. 8   **2.** 7   **3.** 4   **4.** 9   **5.** 8

6. 23   **7.** 34   **8.** 73   **9.** 37   **10.** 216

11. 48   **12.** 59   **13.** 75   **14.** 351   **15.** 483

Page 5   Task M1.11

1. 145 r 3   **2.** 142 r 2   **3.** 39 r 4

4. 452 r 3   **5.** 534 r 8   **6.** 118 r 2

7. 97 r 1   **8.** 47 r 1   **9.** 626 r 4

10. 1811 r 1   **11.** 8   **12.** 32

13. 19   **14.** £171·51

Page 6   Task M1.12/M1.13

1. 28   **2.** 36   **3.** 25   **4.** 16

5. 34   **6.** 43   **7.** 38   **8.** 19

9. 24   **10.** 32 r 28   **11.** B   **12.** 12

13. 16   **14.** £195 more

Page 6 Task M1.14

1. a $5°$ **b** $7°$ **c** $10°$ **d** $15°$

2. $5°$ **3.** $7\,°C$

4. a -5 **b** -5 **c** -3 **d** -11 **e** -1

5. $2°$

Page 7 Task M1.15

1. a 8 **b** 3 **c** -5 **d** -6 **e** -4
   f -6 **g** 0 **h** -9 **i** -8 **j** 0

2. a 8 **b** -3 **c** 6 **d** -6 **e** 6
   f 2

3. owes £22 **4.** B

5. a -5 **b** -3 **c** -2 **d** -9 **e** -5 **f** -10

Page 7 Task M1.16/M1.17

1. a -24 **b** -24 **c** 8 **d** -5 **e** 7
   f 6 **g** 42 **h** -72 **i** -8 **j** -54
   k -5 **l** -5 **m** 9 **n** 56 **o** -27
   p 72 **q** -60 **r** -120

2.

$\times$	-4	-2	-8	9
3	-12	-6	-24	27
-5	20	10	40	-45
6	-24	-12	-48	54
-3	12	6	24	-27

3.

16	$\div$	-4	$\rightarrow$	-4
$\div$		$\times$		
-8	$+$	5	$\rightarrow$	-3
$\downarrow$		$\downarrow$		
-2	$\times$	-20	$\rightarrow$	40

4.

-3	$\times$	-4	$\rightarrow$	12
$\times$		$\times$		
-7	$-$	-6	$\rightarrow$	-1
$\downarrow$		$\downarrow$		
21	$-$	24	$\rightarrow$	-3

5.

-15	$+$	-5	$\rightarrow$	-20
$\div$		$\times$		
-5	$+$	2	$\rightarrow$	-3
$\downarrow$		$\downarrow$		
3	$-$	-10	$\rightarrow$	13

Page 8 Task M1.18/M1.19

1. a 14 **b** 11 **c** 30 **d** 14 **e** 31 **f** 6
   g 36 **h** 25 **i** 10 **j** 3 **k** 10 **l** 10
   m 45 **n** 4 **o** 4 **p** 4

2. a $7 \times (4 + 2) = 42$ **b** $(6 + 9) \div 3 = 5$
   c $(6 + 3) \times 4 = 36$ **d** $(4 + 3) \times (8 - 6) = 14$
   e $(12 + 6) \div 9 = 2$ **f** $(15 - 6) \times (3 + 6) = 81$
   g $8 \times (4 - 2) = 16$ **h** $72 \div (2 + 6) = 9$

3. 5 **4.** 4

5. a 64 **b** 120 **c** 21 **d** 31 **e** 7 **f** 1500
   g 84 **h** 25 **i** 1 **j** 32 **k** 20 **l** 48

Page 8 Task M1.20/M1.21

1. a 25 **b** 49 **c** 36 **d** 1 **e** 900
2. a 4 **b** 6 **c** 10 **d** 8 **e** 1
3. 5 cm **4.** 20
5. a 25 **b** 65 **c** 36 **d** 136 **e** 7
   f 9 **g** 7 **h** 2 **i** 10
6. No **7.** 1 and 4 or 4 and 16 for example
8. a 8 **b** 64 **c** 1 **d** 125 **e** 1000
9. 27
10. a 2 **b** 1 **c** 5 **d** 3
11. 10
12. a 2 **b** 4 **c** 125

Page 9 Task M1.22

1. a 3^4 **b** 2^6 **c** 7^5 **d** 10^3
2. a $9 \times 9 \times 9 \times 9$ **b** $5 \times 5 \times 5 \times 5$
   c $6 \times 6 \times 6 \times 6 \times 6 \times 6$ **d** $2 \times 2 \times 2 \times 2 \times 2 \times 2 \times 2$
3. 2^3 **4.** 5^2

5. a 18   **b** 108   **c** 500

6. 256   **7.** 35

8. a 625   **b** 729   **c** 1024   **d** 100 000

e 16 807   **f** 4096   **g** 279 936   **h** 243

i 4096   **j** 6561   **k** 16 384   **l** 262 144

9. 729, 2187   **10.** 9

Page 10   Task M1.23/M1.24

1. 1, 2, 4, 5, 10, 20   **2.** 1, 2, 3, 4, 6, 12

3. 1, 29   **4.** 1, 2, 3, 6, 9, 18

5. 1, 2, 4, 8, 16, 32   **6.** 1, 2, 5, 10, 25, 50

7. a 2, 6, 14   **b** 5, 9, 11, 13   **c** 2, 5, 11, 13   **d** 2, 6

8. 2, 4, 6, 8, 12, 24   **9.** 5, 7   **10.** 32, 36

11. a 3, 6, 9, 12, 15   **b** 6, 12, 18, 24, 30

c 9, 18, 27, 36, 45   **d** 8, 16, 24, 32, 40

e 12, 24, 36, 48,60

12. a 18, 42, 54   **b** 21, 28, 42, 63   **c** 21, 63

13. 6, 12, 18, 24, 30; 10, 20, 30, 40, 50; 30

14. a 40   **b** 24   **c** 60

15. 64, 72   **16.** 35 mins   **17.** 72

Page 11   Task M1.25

1. a 1, 2, 3, 6, 9, 18   **b** 1, 2, 3, 5, 6, 10, 15, 30   **c** 6

2. a 1, 2, 3, 5, 6, 10, 15, 30   **b** 1, 3, 5, 9, 15, 45   **c** 15

3. a 10   **b** 5   **c** 12   **d** 8

4. Sophia is correct   **5.** 1, 2, 3, 6

Page 12   Task M1.26

1. a 45   **b** 28   **c** 36

2. a 7   **b** 5   **c** 7

3. a $3 \times 2 \times 2 \times 2 \times 2$ **b** $2 \times 3 \times 2 \times 5$

4. a 3×5^2 **b** $2^2 \times 11$ **c** $2^4 \times 5$ **d** $2 \times 3^3 \times 11$

5. 21

6. $3^2 \times 5 \times 7$; $3^2 \times 5 \times 11$; $3^2 \times 5$ (45)

7. $2^2 \times 3^2 \times 11$; $2^2 \times 3 \times 5 \times 7$; $2^2 \times 3$ (12)

Page 13   Task E1.1

1. a $6 \cdot 3 \times 10^4$ **b** $5 \cdot 96 \times 10^3$ **c** $9 \cdot 6 \times 10^{-2}$

d $5 \cdot 8 \times 10^{-4}$ **e** $7 \cdot 8 \times 10^6$ **f** 4×10^{-1}

2. a 3×10^3 **b** 7×10^4 **c** $3 \cdot 4 \times 10^2$

d $8 \cdot 9 \times 10^4$ **e** 4×10^{-3} **f** 7×10^{-4}

g 9×10^{-1} **h** $1 \cdot 8 \times 10^{-3}$

3. should be $3 \cdot 7 \times 10^3$ **4.** $2 \cdot 8 \times 10^4$

5. a 60 000   **b** 300   **c** 0·03   **d** 56 000

e 240 000   **f** 0·0086   **g** 4160   **h** 0·768

6. $4 \cdot 7 \times 10^2$ **7.** $6 \cdot 5 \times 10^{-2}$

8. a 7×10^{-4} **b** $5 \cdot 3 \times 10^4$ **c** $9 \cdot 6 \times 10^{-2}$

d $4 \cdot 87 \times 10^{-1}$ **e** $4 \cdot 9 \times 10^7$ **f** $5 \cdot 76 \times 10^5$

g $7 \cdot 4 \times 10^{-4}$ **h** $8 \cdot 24 \times 10$ **i** 1×10^{-1}

j $8 \cdot 64 \times 10^{-7}$ **k** $6 \cdot 18 \times 10^6$ **l** $4 \cdot 2 \times 10^7$

Page 13   Task E1.2

1. a $1 \cdot 82 \times 10^{19}$ **b** $4 \cdot 05 \times 10^{12}$

c $7 \cdot 6 \times 10^{25}$ **d** 5×10^{14}

2. $2 \cdot 8 \times 10^{18}$ **3.** $3 \cdot 007 \times 10^8$ **4.** $1 \cdot 96 \times 10^{15}$

5. a Brooke, Carter, Brianna, Matt, Vanya

b £700   **c** £2768·80

6. a $3 \cdot 6 \times 10^{-15}$ **b** 2×10^{21} **c** 2×10^{-16}

d $8 \cdot 3 \times 10^{-15}$ **e** $3 \cdot 75 \times 10^{36}$ **f** 8×10^{26}

7. $7 \cdot 48 \times 10^{-20}$ g   **8.** £($1 \cdot 62 \times 10^{12}$)   **9.** 5×10^{10}

Page 15   Task M2.1

1. a $2n$ **b** $n + 15$ **c** $4n + 15$

2. a $49m$ **b** $54n$ **c** $49m + 54n$

3. $5x + 3y + 2m$

4. a $n - y$ **b** $m + n + 62$ **c** $x - 3$ **d** $m - 62$

5. a $16n + 9y$ **b** $16n + 4 \cdot 5y$ **6.** $\frac{1}{2}(15n - m)$

Page 16   Task M2.2

1. 12	2. 21	3. 27	4. 43	5. 10
6. 16	7. 1	8. 49	9. 58	10. 8
11. 90	12. 90	13. 63	14. 6	15. 33
16. 2	17. 6	18. 74	19. 84	20. 17
21. 10	22. 63	23. 48		

4 Answers

Page 16 Task E2.1

1. -8	**2.** -3	**3.** -5	**4.** 4	**5.** 14
6. 17	**7.** 11	**8.** 16	**9.** 41	**10.** 20
11. 0	**12.** 19	**13.** -14	**14.** 12	**15.** 42
16. 16	**17.** 11	**18.** 12	**19.** -7	**20.** 32
21. 0	**22.** 80	**23.** 45		

Page 16 Task M2.3

1. 17	**2.** 41	**3.** 2	**4.** 54	**5.** 105
6. 8	**7.** 250	**8.** 60	**9.** 18	**10.** $42 \cdot 408$

Page 17 Task E2.2

1. a 11　　**b** 114　　**c** 45

2. a 46　　**b** 150

3. a 65　　**b** 7　　**c** 23　　**d** 20

4. a 108　　**b** 300　　**c** 768

5. $1 \cdot 35 \times 10^{18}$

6. a -75　　**b** -17

7. $46 \cdot 5$

Page 18 Task M2.4

1. $9x + 9y$　　　**2.** $8p + 6q$　　　**3.** $13a + 6b$

4. $2m + 9p$　　　**5.** $4a + 5b$　　　**6.** $2f + 6g$

7. $3a + 5b$　　　**8.** $5p + q$　　　**9.** $2m + 3q$

10. $a + b + 9q$　　**11.** $x + 5y$　　　**12.** $2m + p + $

13. a

b

c

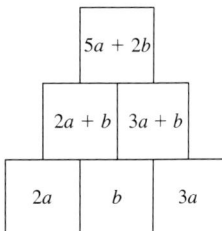

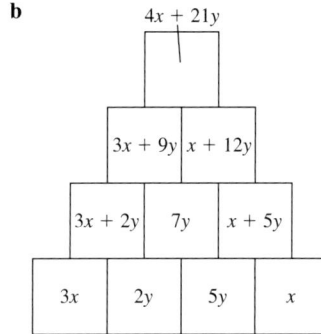

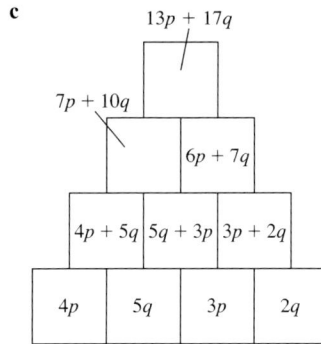

Page 19 Task M2.5

1. $-3b$	**2.** $5x$	**3.** $-2y$
4. $9q - 3p$	**5.** $6b - 2a$	**6.** $8x + 2$
7. $3 - 3c$	**8.** $6 - 2m$	**9.** $9f + 4$
10. $15a^2$	**11.** $3x^2$	**12.** $8ab$
13. $6xy + 3x$	**14.** $12mn + 4$	**15.** $6p^2 - p$
16. $9m^2 - 3m$	**17.** $5a + ab + b$	**18.** $7a^2 + 2ab$

19. Bella is correct

22. a $2pq + 5$　　**b** $6pq + 8p + 7$　　**c** $2pq + 8p + 9$

23. £$(n^2 + n)$　　**24.** $2n + 8$　　**25.** $3a^2 + 3a + 9ab$

Page 20 Task M2.6

1. $9a$	**2.** $8b$	**3.** $21m$	**4.** $35n$
5. $48y$	**6.** $8x$	**7.** $9y$	**8.** $6a$
9. $3b$	**10.** $40m$	**11.** n^2	**12.** y^2
13. $6a^2$	**14.** $9f^2$	**15.** $7q^2$	**16.** $6y^2$
17. $15b^2$	**18.** $32mp$	**19.** $45ab$	**20.** $100a^2$
21. $6m$	**22.** $5x$	**23.** $5a$	**24.** $18n^2$

25. correct　　**26.** $\dfrac{2n}{n} = 2$

Page 20   Task M2.7

1. F   2. F   3. T   4. F

5. T   6. T   7. F   8. T

9. T   10. $-24ab$ 11. $6mp$ 12. $2a$

13. $-3x$ 14. $10a^2$ 15. $36fg$ 16. $-21pq$

17. $-32ac$ 18. $-3a^2$ 19. $-66y$ 20. $-18b^2$

21. $21a^2$ 22. $5a \times 3b = 15ab$

23. $9m \times 4m = 36m^2$ 24. $\dfrac{30n}{10} = 3n$

25. $6x \times 4y = 24xy$ 26. $\dfrac{12b}{4} = 3b$

27. $4x - 8y = -32y$ 28. $\dfrac{-42y}{7} = -6y$

29. $\dfrac{-40n}{-8} = 5n$ 30. $-5p \times -9q = 45pq$

31. $48mn$

Page 21   Task M2.8

1. a 5^7 b 6^2 c 3^5 d 9^6 e 2^8 f 4^8
   g 2^9 h 5^7 i 5^4

2. a 2^7 b 3^5 c 6^4 d 5^5 e 8 f 4^8

3. a F   b F   c T

4. a 3^3 b 5^2 c 9^3

5. $3^2 \times 2^3 = 9 \times 8 = 72$ but $6^5 = 7776$

6. $\dfrac{3^2 \times 3^6}{3^3}$ 7. $2^5 = 32$

Page 21   Task E2.3

1. a 4^6 b 2^9 c 7^8 d 5^{11} e 6^6 f 3^4
   g 4^{15} h 2^3 i 7^5

2. 1

3. a x^7 b y^9 c a^4 d m^4 e x^8 f 1
   g y^{15} h 1 i x^4

4. a F   b F   c F   d T   e T   f F

5. A

6. a n b x^3 c a^4 d 1 e m^2 f n^2
   g x^4 h m i x^3

7. a T   b F   c T   d T

8. a $6n^9$ b $9m^8$ c $28n^8$ d $7m^2$ e $3n^4$ f $15m^5$

Page 22   Task M2.9

1. $3a + 12$ 2. $a^2 + ab$ 3. $5m + 10$

4 $4x - 12$ 5. $6a - 48$ 6. $6y + 10$

7. $18m - 36$ 8. $3x + 3y$ 9. $12a - 6b$

10. $5m + 15p$ 11. $14x + 35$ 12. $12p - 16q$

13. $ab + ac$ 14. $x^2 - xy$ 15. $m^2 - 3mp$

16. $2cd + c$ 17. $2p^2 + 2pq$ 18. $10a + 5b$

19. $4ab - a$ 20. $m^2 + 8mp$

21. $m(m - n) = m^2 - mn = 12$

22. Austin correct

Page 23   Task E2.4

1. $-4x - 28$ 2. $-18b + 6$ 3. $-3a - 6$

4. $-6b + 24$ 5. $-5x + 15$ 6. $-6m + 8$

7. $-ab + ac$ 8. $-2m + mp$ 9. $-xy - yz$

10. $-x^2 - 3xy$ 11. $-a + b$ 12. $-p - q$

13. $-2ab + 3b$ 14. $-5fg - 2hf$ 15. $-q^2 + 8qr$

16. $9a^2 + 12ab$ 17. $-32x^2 + 24xy$

18. $-4(m - 2) = -4m + 8$ 19. $-5(n + 3) = -5n - 15$

20. $-a(a + 4) = -a^2 - 4a$ 21. $-p(p - 4q) = -p^2 + 4pq$

22. $-3m(4m + 7n) = -12m^2 - 21mn$

23. $-7y(2y - 2p) = -14y^2 + 14py$

Page 24   Task M2.10

1. $7a + 21$ 2. $26x + 15$ 3. $3a + 19$

4 $6m + 15$ 5. $8x + 30$ 6. $8y + 26$

7. $22b + 36$ 8. $25a + 42$ 9. $7x + 20$

10. $13p + 21$ 11. $24m + 33$ 12. $25a + 6$

13. $28y + 8$ 14. $5n + 12x + 27$ 15. $31b + 30$

16. $34c + 25$ 17. $20n + 45$ 18. $3a^2 + ab$

19. $7mn + 5n^2$ 20. $4a^2 + 17ab$

Page 25   Task M2.11

1. 3   2. 1   3. 3

4 $3b$ 5. $4y + 7$ 6. $5x - 3$

7. $2(3x + 5)$ 8. $4(2a + 3)$ 9. $10(p - 4)$

10. $5(4y - 5)$ 11. $3(4m + 3)$ 12. $12(3b - 1)$

13. $3(3x + 2y)$ 14. $4(4a + 3b)$ 15. $4(6m - 5p)$

16. $5(9f + 7g)$ 17. $3(7a - 5b)$ 18. $10(3x - 5y)$

19. $2(4p + 3q - 5r)$ 20. $5(3x - 6y - 4z)$ 21. $7(5a - 3b + 7c)$

Page 25   Task M2.12

1. f **2.** y **3.** 5

4 7   **5.** f **6.** $2a + 9c$

7. $y(x + z)$ **8.** $a(a - 6)$ **9.** $b(b + 4)$

10. $c(c + 9)$ **11.** $p(m - q)$ **12.** $3x(y + 3z)$

13. $5a(2b - 3c)$ **14.** $3w(6z - 5y)$ **15.** $3f(4g + 7)$

16. $2a(2a - 3)$ **17.** $5p(p - 6q)$ **18.** $6m(3p + 5)$

19. $4q(2p - 5q)$ **20.** $4y(4xz - 7y)$ **21.** $11a(3a + 5bc)$

22. $6m(m - 2n)$ **23.** $4x(2x + 3y)$ **24.** $6n(5m - 4n)$

25. $4y(5y - 3xz)$

Page 26   Task M2.13

1. a 5   **b** 7   **c** 9   **d** 16   **e** 31   **f** 28

g 42   **h** 23   **i** 15   **j** 52   **k** 77   **l** 15

2. 39

3. a 5   **b** 4   **c** 7   **d** 35   **e** 15   **f** 8

g 60   **h** 28   **i** 40   **j** 6   **k** 62   **l** 6

4. 18   **5.** 12%

6. a 41   **b** 7   **c** 42   **d** 10   **e** 37   **f** 4

g 8   **h** 81

Page 26   Task E2.5

1. a -2 **b** -1 **c** -2 **d** -3 **e** -8 **f** -2

g -0.8 **h** 4   **i** -7 **j** -12 **k** -25 **l** -21

2. -5 **3.** $4\frac{1}{2}$

4. a $\frac{1}{2}$ **b** $\frac{5}{2}$ **c** $\frac{7}{3}$ **d** $\frac{4}{5}$ **e** $-\frac{3}{2}$ **f** $\frac{7}{5}$

g $-\frac{4}{7}$ **h** $-\frac{11}{3}$

5. $\frac{10}{3} = 3\frac{1}{3}$

6. a 20   **b** -18 **c** -20 **d** -48 **e** $-\frac{3}{5}$ **f** -14

g $-\frac{7}{9}$ **h** $-\frac{5}{6}$ **i** $\frac{2}{7}$ **j** -6 **k** $-\frac{5}{8}$ **l** $-\frac{3}{11}$

Page 27   Task M2.14

1. $n = 6$ **2.** $5n = 20, n = 4$ **3.** $4x = 8, x = 2$

4. 4   **5.** 3   **6.** 2   **7.** 5   **8.** 3

9. 4   **10.** 3   **11.** 5   **12.** 9   **13.** 7

14. 6   **15.** 7   **16.** 7   **17.** 9   **18.** 7

19. 10   **20.** 8   **21.** 7   **22.** 4

Page 28   Task E2.6

1. $n = \frac{2}{3}$ **2.** $5n = 3, n = \frac{3}{5}$ **3.** $-8 = 2n, -4 = n$

4. $\frac{3}{5}$ **5.** $\frac{5}{2}$ **6.** $\frac{3}{4}$ **7.** $\frac{5}{8}$ **8.** $\frac{7}{3}$

9. $\frac{11}{2}$ **10.** $\frac{3}{4}$ **11.** $\frac{7}{9}$ **12.** 2.3 **13.** $\frac{1}{5}$

14. $\frac{5}{7}$ **15.** $-\frac{2}{3}$ **16.** $-\frac{1}{6}$ **17.** $-\frac{1}{6}$ **18.** -3

19. -4 **20.** $\frac{-3}{8}$ **21.** 8   **22.** -5 **23.** $-\frac{5}{6}$

Page 28   Task M3.1/M3.2

2. a 3   **b** 8   **c** 10   **d** 50

3. a 2   **b** 5   **c** 9   **d** 20

4. 7   **5.** 12   **6.** 21   **7.** 30   **8.** 12   **9.** 45

10. a $\frac{1}{2}$ **b** $\frac{7}{18}$ **11. a** $\frac{1}{3}$ **b** $\frac{2}{3}$

12. a $\frac{16}{41}$ **b** $\frac{25}{41}$ **13.** $\frac{1}{4}$

14. a $\frac{1}{6}$ **b** $\frac{1}{3}$ **c** $\frac{3}{4}$ **d** $\frac{5}{6}$ **e** $\frac{17}{60}$ **f** $\frac{3}{5}$

Page 29   Task M3.3

1. 45   **2.** 15   **3.** 16   **4.** 18   **5.** 15

6. 56   **7.** 21   **8.** 72   **9.** £15   **10.** 368 g

11. 24   **12.** 48   **13.** 700 g   **14.** £6   **15.** £224

16. 120 cm³

Page 30   Task M3.4/M3.5

1. A and B

3. a $\frac{15}{20}$ **b** $\frac{4}{12}$ **c** $\frac{10}{16}$ **d** $\frac{8}{36}$

e $\frac{15}{40}$ **f** $\frac{35}{100}$ **g** $\frac{24}{30}$ **h** $\frac{45}{81}$

4. a $\frac{9}{10}$ **b** $\frac{2}{5}$ **c** $\frac{2}{5}$ **d** $\frac{3}{4}$ **e** $\frac{1}{3}$

f $\frac{1}{4}$ **g** $\frac{3}{4}$ **h** $\frac{2}{3}$ **i** $\frac{7}{9}$ **j** $\frac{8}{11}$

5. a, d, f   **6.** RADISH

Page 31   Task M3.6

2. a $\dfrac{8}{16}$ **b** $\dfrac{9}{16}$ **3. a** $\dfrac{5}{8}$ **b** $\dfrac{26}{30}$ **c** $\dfrac{6}{7}$

4. a $\dfrac{7}{20}, \dfrac{3}{10}, \dfrac{1}{4}$ **b** $\dfrac{11}{16}, \dfrac{5}{8}, \dfrac{19}{32}$ **c** $\dfrac{13}{18}, \dfrac{2}{3}, \dfrac{5}{9}$ **d** $\dfrac{1}{6}, \dfrac{1}{8}, \dfrac{5}{48}$

5. a $\dfrac{6}{16}$ **b** $\dfrac{8}{16}$ **c** YES

6. a $\dfrac{10}{40}$ **b** $\dfrac{12}{40}$ **c** $\dfrac{11}{40}$

7. a $\dfrac{25}{35}$ **b** $\dfrac{28}{35}$ **c** $\dfrac{26}{35}$ or $\dfrac{27}{35}$

8. a $\dfrac{9}{30}$ **b** $\dfrac{21}{24}$ **c** $\dfrac{15}{20}$

Page 32   Task M3.7

1. Yes   **2.** Yes   **3.** No

4. a $\dfrac{3}{100}$ **b** $\dfrac{41}{50}$ **c** $\dfrac{2}{5}$ **d** $\dfrac{13}{250}$ **e** $\dfrac{3}{20}$

5. a $\dfrac{55}{100} = 0{\cdot}55$ **b** $\dfrac{35}{1000} = 0{\cdot}035$

6. a $0{\cdot}15$ **b** $0{\cdot}76$ **c** $0{\cdot}515$ **d** $0{\cdot}375$ **e** $0{\cdot}52$

7. not correct $\left(\dfrac{2}{5} = 0{\cdot}4\right)$ **8.** $\dfrac{7}{20}$ **9.** $0{\cdot}11$

10. a $0{\cdot}5$ **b** $0{\cdot}09$ **c** $0{\cdot}416$

Page 33   Task M3.8

1. a $0{\cdot}5$ **b** $0{\cdot}04$ **c** $0{\cdot}742$

2. True   **3.** False

4. a D   **b** D, A, C, B

5. correct

6. a $0{\cdot}003, 0{\cdot}03, 0{\cdot}3$ **b** $0{\cdot}091, 0{\cdot}902, 0{\cdot}91, 0{\cdot}92$

   c $0{\cdot}07, 0{\cdot}073, 0{\cdot}712, 0{\cdot}75$ **d** $0{\cdot}048, 0{\cdot}408, 0{\cdot}418, 0{\cdot}48$

   e $7{\cdot}06, 7{\cdot}07, 7{\cdot}1, 7{\cdot}102, 7{\cdot}13$

7. Luka, David, Nathan

8. a F   **b** T   **c** T   **d** F   **e** T   **f** F

9. $\dfrac{3}{25}, \dfrac{3}{20}, 0{\cdot}18, 0{\cdot}2, 0{\cdot}23, \dfrac{1}{4}$

Page 34   Task M3.9/3.10

1. $2\dfrac{2}{3}$ **2.** $1\dfrac{1}{6}$ **3.** $2\dfrac{4}{5}$ **4.** $1\dfrac{3}{5}$

5. $2\dfrac{1}{3}$ **6** $2\dfrac{1}{8}$ **7.** $4\dfrac{1}{2}$ **8.** $4\dfrac{1}{3}$

9. $3\dfrac{4}{5}$ **10.** $7\dfrac{3}{4}$ **11.** $2\dfrac{7}{8}$ **12.** $4\dfrac{4}{9}$

13. $\dfrac{7}{5}$ **14.** $\dfrac{21}{8}$ **15.** $\dfrac{23}{6}$ **16.** $\dfrac{19}{5}$

17. $\dfrac{14}{3}$ **18.** $\dfrac{23}{4}$ **19.** $\dfrac{23}{8}$ **20.** $\dfrac{29}{6}$

21. $\dfrac{16}{5}$ **22.** $\dfrac{39}{4}$ **23.** $\dfrac{26}{3}$ **24.** $\dfrac{75}{8}$

25. $\dfrac{19}{3} = 6\dfrac{1}{3}$ **26.** $5\dfrac{7}{8}$ **27.** $8\dfrac{3}{4}$ **28.** $5\dfrac{1}{5}$

29. $\dfrac{20}{3}$ **30.** $\dfrac{15}{7}$ **31.** $9\dfrac{1}{9}$

32. a $\dfrac{23}{7}$ **b** $\dfrac{44}{9}$ **c** $6{\cdot}5$

Page 35   Task M3.11

1. $\dfrac{3}{4}$ **2.** $\dfrac{1}{9}$ **3.** $\dfrac{3}{11}$ **4.** $\dfrac{11}{20}$

5. $\dfrac{5}{7}$ **6.** $\dfrac{8}{9}$

7. a $\dfrac{37}{56}$ **b** $\dfrac{7}{12}$ **c** $\dfrac{43}{90}$

8. part a   **9.** c   **10.** $\dfrac{38}{45}$ **11.** $\dfrac{9}{35}$

12. $\dfrac{7}{24}$ **13.** $\dfrac{7}{10}$ **14.** $\dfrac{2}{15}$ **15.** $\dfrac{5}{12}$

16. $\dfrac{2}{5}$ **17.** $\dfrac{53}{110}$ **18.** $\dfrac{13}{30}$ **19.** $\dfrac{3}{20}$

Page 36   Task E3.1

1. a $\dfrac{13}{5}$ **b** $3\dfrac{7}{20}$

2. a $7\dfrac{1}{6}$ **b** $6\dfrac{1}{2}$ **c** $8\dfrac{9}{20}$ **d** $7\dfrac{17}{24}$

3. $8\dfrac{11}{12}$ km   **4.** b   **5.** $\dfrac{13}{20}$ metre

6. $2\dfrac{1}{4}$ in middle row, $\dfrac{3}{4}$ in bottom row

7. $\dfrac{1}{6}$ **8.** $\dfrac{9}{10}$ hectares

Page 37   Task M3.12

1. $\dfrac{1}{6}$ **2.** $\dfrac{1}{15}$ **3.** $\dfrac{3}{8}$ **4.** $\dfrac{2}{5}$

5. $\dfrac{1}{32}$ **6.** $\dfrac{1}{42}$ **7.** $\dfrac{1}{14}$ **8.** $\dfrac{2}{15}$

9. $4\dfrac{1}{2}$ **10.** $3\dfrac{1}{3}$ **11.** $6\dfrac{1}{4}$ **12.** 4

13. Tony runs for 1 minute longer

14. $\dfrac{5}{24}$ **15.** B   **16.** Callum

Page 38   Task E3.2 _____

1. $\frac{2}{3} \times 2\frac{1}{5} = \frac{2}{3} \times \frac{11}{5} = \frac{22}{15} = 1\frac{7}{15}$

2. $3\frac{1}{2} \times 4\frac{1}{3} = \frac{7}{2} \times \frac{13}{3} = \frac{91}{6} = 15\frac{1}{6}$

3. $A = 6\frac{3}{10}, B = 3\frac{17}{21}, C = 7\frac{13}{20}$

4. **a** $30\,m^2$ **b** $21\frac{3}{4}\,m^2$ **c** $38\frac{1}{2}\,m^2$

5. **a** $\frac{3}{16}$ **b** $\frac{13}{27}$ 6. $\frac{11}{12}$ 7. $4\frac{2}{5}$

Page 38   Task M3.13 _____

1. $\frac{1}{7} \div \frac{1}{3} = \frac{1}{7} \times \frac{3}{1} = \frac{3}{7}$ 2. $\frac{4}{9} \div \frac{3}{5} = \frac{4}{9} \times \frac{5}{3} = \frac{20}{27}$

3. 20   4. 18   5. $\frac{4}{7}$ 6. $1\frac{1}{2}$ 7. $\frac{8}{9}$ 8. $\frac{9}{14}$

9. $\frac{9}{44}$ 10. $\frac{3}{5}$ 11. 4   12. A   13. 33

14. $\frac{3}{8} \div \frac{2}{3}$ larger by $\frac{2}{16}\left(\frac{1}{8}\right)$ 15. 8

Page 39   Task E3.3 _____

1. $2\frac{1}{4} \div \frac{3}{5} = \frac{9}{4} \div \frac{3}{5} = \frac{9}{4} \times \frac{5}{3} = \frac{45}{12} = \frac{15}{4} = 3\frac{3}{4}$

2. $1\frac{1}{2} \div 1\frac{1}{6} = \frac{3}{2} \div \frac{7}{6} = \frac{3}{2} \times \frac{6}{7} = \frac{18}{14} = \frac{9}{7} = 1\frac{2}{7}$

3. $4\frac{3}{8}$ 4. $2\frac{5}{8}$ 5. $2\frac{8}{35}$ 6. $2\frac{2}{3}$

7. $A \rightarrow P, B \rightarrow S, C \rightarrow Q$ 8.   28

5. **a** $3\frac{5}{12}$ **b** $\frac{9}{20}$ 10. 7

Page 40   Task E3.4 _____

1. **a** $\frac{1}{3}$ **b** $0\cdot1$ **c** $0\cdot25$ 2. $2\cdot5$

3. **a** $\frac{1}{7}$ **b** $\frac{1}{9}$ **c** 5   **d** $\frac{1}{20}$ **e** 6   **f** 100

4. $\frac{3}{5}$ 5. $\frac{9}{2}$ 6. $\frac{11}{8}$

7. **a** $\frac{4}{3}$ **b** $\frac{10}{7}$ **c** $\frac{19}{4}$ **d** $\frac{2}{7}$ **e** $\frac{n}{m}$

8. $\frac{1}{n}$ 9. 12   10. 2 and $0\cdot5$, 15 and $\frac{1}{15}$, $\frac{2}{3}$ and $\frac{3}{2}$

Page 41   Task M4.1 _____

1. **a** Ac   **b** Ac   **c** Re   **d** Ob   **e** Ac
   f Ob   **g** Re   **h** Re   **i** Ac   **j** Re

2. **a** $90°$ **b** $90° < Ob < 180°$ **c** $180°$

3. $120°, 138°$ 5. No

6. **a** no   **b** $Q\hat{R}S$ **c** $T\hat{P}Q$ **d** $R\hat{S}T$ or $P\hat{T}S$
   e pentagon

7. **a** 2   8. **a** Ac   **b** Ac

9. **a** $P\hat{N}M$ **b** $P\hat{S}R$ **c** $G\hat{F}H$

Page 43   Task M4.2 _____

1. $a = 150°$ 2. $b = 63°$ 3. $c = 40°$ 4. $d = 135°$

5. $e = 156°$ 6. $f = 37°$ 7. $g = 162°$ 8. $54°$

9. **a** $68°$ 10. **a** $53°$

Page 44   Task M4.3 _____

1. **b** $90°$ 3. $a = 75°$ 4. $b = 66°$ 5. $c = 49$

6. $d = 112°, e = 68°$ 7. $f = 63°, g = 117°$

8. $h = 53°, i = 45°$ 9. $j = 34°, k = 56°$

10. $l = 69°, n = 111°, m = 69°$

11. **a** $90°$ **b** $41°$

12. **a** $61°$ **b** $61°$ **c** $33°$

Page 45   Task M4.4 _____

1. $a = 70°$ 2. $b = 48°, c = 84°$

3. $d = 65°, e = 50°, f = 115°$ 4. $g = 80°, h = 80°$

5. $i = 60°, j = 120°$ 6. $k = 52°, l = 52°, m = 52°$

7. $n = 72°, o = 54°, p = 54°$ 8. $q = 57°, r = 57°, s = 66°$

9. **a** $110°$ **b** $40°$

11. $P\hat{R}Q = 38°$ and $P\hat{Q}R = 104°$

12. $30°$

Page 46   Task M4.5 _____

1. *a* 107°   **2.** *b* 39°   **3.** *c* 47°, *d* 47°

4. *e* 110°, *f* 70°, *g* 110°

5. *h* 124°, *i* 124°, *j* 56°, *k* 143°

6. *l* 86°, *m* 94°, *n* 54°, *o* 54°

7. *p* 115°, *q* 115°, *r* 65°, *s* 115°

8. *t* 86°, *u* 64°, *v* 30°

9. CB̂D, CÂE or BD̂C, CÊA

10. a 110° (corresponding to DF̂I)

   b 75° (alternate to AĈI)

   c 70° (different reasons may be given)

12. yes

Page 47   Task M4.6/M4.7 _____

4. 2   **5.** 4   **6.** 2   **7.** 5

8. 3   **9.** 1   **10.** 8   **11.** 6

Page 49   Task M4.8 _____

2. None   **4.** Trapezium   **5.** 90°

7. equal, parallel, half   **8.** 1

Page 49   Task M4.9 _____

1. *a* 60°   **2.** *b* 103°

3. *c* 77°, *d* 103°   **4.** *e* 75°, *f* 105°

5. *g* 63°, *h* 84°   **6.** *i* 72°, *j* 72°, *k* 58°

7. *l* 59°, *m* 66°   **8.** 108°   **9.** 104°

10. 125°   **11.** 55°   **12.** 45°, 80°, 100°, 135°

Page 51   Task M4.10 _____

1. 360°   **2.** 45°   **3. a** 36°   **b** 144°

4. 140°   **5. a** 20°   **b** 15°   **c** 8°

6. a 160°   **b** 165°   **c** 172°

7. 15   **8.** 20   **9.** 162°

Page 52   Task E4.1 _____

1. 3 triangles, 540°   **2.** 1080°   **3.** 2340°

4. 4 triangles, 720°, 637°, *x* = 83°

5. *a* 20°   **6.** *b* 53°   **3.** *c* 115°

8. 135°   **9.** 8° in error

Page 53   Task M4.11 _____

1. C, E   **2.** Q, T   **3.** A, E

1. a I, K   **b** C, L   **c** G, N   **d** H, M

Page 54   Task M4.12 _____

1. a HELP ME OUT

   b (3, 1)(3, 4)(1, 0)(3, 1)(4, 3)(0, 4)(1, 3)(3, 4)

Page 56   Task M4.13 _____

1. a $\binom{3}{2}$ **b** $\binom{2}{3}$ **c** $\binom{-2}{-1}$ **d** $\binom{3}{0}$

   e $\binom{4}{-4}$ **f** $\binom{-3}{1}$ **g** $\binom{0}{5}$ **h** $\binom{-3}{-1}$

   i $\binom{5}{4}$ **j** $\binom{2}{5}$ **2. e** $\binom{3}{-2}$

Page 58   Task M4.15 _____

1. d reflect in the *y*-axis

2. new vertices should be $(0, -1)$, $(1, -2)$, $(3, -1)$

3. stays the same

Page 59   Task M4.16 _____

9. d different position or diagonally to side of each other, etc.

10. g 90° anticlockwise centre $(-3, 1)$

Page 60   Task M4.17 _____

1. a 3   **b** not an enl.

5. b 4   **d** 36   **e** 9 times bigger (i.e. (scale factor)²)

Page 61   Task M5.1 _____

1. a 5·6   **b** 43·99   **c** 61·24   **d** 70·33

2 A, B, D   **3.** B

4. a £5·19   **b** £7·36   **c** £1·79   **d** £3·28

5. £164·56

6. a 6·17   **b** 9·03   **c** 33·8   **d** 13·6

   e 32·91   **f** 14·2

7. 133·14   **8.** £11·92   **9.** B   **10.** 14·99 km

11. £2·15   **12.** 27·27   **13.** 42·23 cm   **14.** 3·82 secs

Page 62   Task M5.2

1. a T   b T   c T   d T
2. a 2·1   b −0·35 c 0·27   d 0·054 e 0·56 f 0·16
3. a £5·28   b £14·56   c £57·04
4. £13·12   5. 343·5 kg   6. A
7. a 2·36   b −1·82 c 1·02   d 0·085 e 0·104 f −2·259
8. 6 packets for £25·08
9. a 0·6   b −0·29 c 0·04   d 1·3   e 0·071 f 0·48
   g −6·31 h 0·08
10. €25·20
11. a 0·1   b 0·01   c −0·01 d 0·01   e 6   f 34

Page 63   Task M5.3

1. a 6·2   b 2·14   c 8·05   d 5·6   e 9·5   f 1·625
2. a 2·38   b 5·75   c 13·75 d 1·125
3. £15·65   4. £1·08
5. a 1·9   b 6·14   c −5·24   d 8·62
   e 0·69   f −0·026   g −3·325   h 0·0312
   i −2·34
6. Multipack by 6p per tin
7. 2 packs of 8 are 48p cheaper

Page 64   Task M5.4

1. b, c, e, f
2. a 4·81   b 0·36   c 28·19   d 5·65
3. 10 750   4. 6·742, 6·735 12, 6·738
5. 74 500   6. 7·48
7. a 11·2   b 2·09   c 9·2   d 0·15
   e 16·981   f 5·39   g 13·469   h 0·175

Page 65   Task M5.5

1. a, c   2. 45 cm²   3. £180   4. £600
5. A → R, B → T, C → S, D → P, E → Q
6. a 17·296   b 47   c 4·7

Page 66   Task M5.6

1. a 3   b 23·62   c 137·7
   d 5·58   e 11·5   f 6·9169
2. a £16·60   b £34·20   c £13·56

d £4·80   e £91·20   f £37·80
3. a 4   b 2   c 0·2   d 3·75
4. A → R, B → U, C → S, D → P, E → Q, F → T

Page 66   Task M5.7

1. a $\frac{2}{15}$ b $1\frac{1}{3}$ c $3\frac{33}{35}$ d $2\frac{5}{11}$
2. a 11·56   b 0·7   c 3·7   d 256
   e 512   f 6·2   g 1024   h 9
3. a 0·4   b 0·55   c 0·95
   d 3·25   e 2·1   f 5·7
4. a 3·51   b 32·63   c 55·36
   d 12·08   e 5·58   f 4·14
   g 3·72   h 17·97   i 0·85
5. a $\frac{5}{9} \times \frac{4}{7} = \frac{20}{63}$ b $3\frac{1}{2} - 1\frac{2}{3} = 1\frac{5}{6}$ c $\frac{2}{3} \div \frac{4}{7} = 1\frac{1}{6}$
   d $\frac{3}{8} + 2\frac{1}{3} = 2\frac{17}{24}$ e $4\frac{1}{2} \times 3\frac{1}{5} = 14\frac{2}{5}$ f $2\frac{3}{4} \div 1\frac{1}{2} = 1\frac{5}{6}$

Page 67   Task M5.8

1. a 3·17   b 5·62   c 41·7   d 0·147
   e 16·6   f 61 700   g 26 400   h 318
2. 37 169, 37 212, 37 241
3. a, c, d   4. 0·0745
5. a 17·3   b 240   c 280   d 213
   e 35·4   f 84 000   g 80   h 0·55

Page 68   Task M5.9

1. a 70   b 4   c 400   d 10   e 5
   f £3000   g £60 000 h 3000   i 5
2. a 384   b 27   c 1036·8
3. a 486   b 714·42   c 71·442
4. a 37 107   b 63   c 589
5. eg. ≤4 × 30 ≤ 120 so not correct
6. eg. $\frac{4·2 \times 2}{0·5 \times 2} = \frac{8·4}{1} = 8·4$ so correct

Page 69   Task M5.10

1. 73%   2. 55%   3. 32%
4. a 42%   b 58%   5. 47%

6. a $\frac{17}{100}$ **b** $\frac{1}{5}$ **c** $\frac{29}{100}$ **d** $\frac{3}{25}$ **e** $\frac{1}{4}$

f $\frac{3}{5}$ **g** $\frac{6}{25}$ **h** $\frac{7}{20}$ **i** $\frac{22}{25}$ **j** $\frac{8}{25}$

k $\frac{9}{100}$ **l** $\frac{19}{20}$

7. $\frac{1}{50}$

8. a 70% **b** 40% **c** 12% **d** 76%

e 65% **f** 32%

9. Luka 92% (higher than 90% and 88%)

10. 85%

Page 70   Task M5.11

1. 95%

2. a 36% **b** 64%

3. a 38% **b** 55% **c** 60% **4.** 80%

5. a 30% **b** 24% **c** 8% **d** 18%

6. 65% **7.** 22% **8.** 53% **9.** Tamsin by 2%

10. a 9% **b** 53% **c** 2% **d** 5%

11. 1·67% **12.** 4%

Page 71   Task M5.12

1. £13 **2. a**

3. a £44 **b** £56 **c** £30 **d** £49

4. £357 **5.** £992

6. a £560 **b** £360 **c** £51

7. £23 280 **8.** £940

9. a £72 **b** £432

10. Quote [2] cheaper by £700

11. £1800

Page 72   Task M5.13

1. 368

2. a £3·15 **b** £15·39 **c** £6·87 **d** £3·98

3. £14 570 **4.** £7·02

5. a £67·90 **b** £69·36 **c** £142·56 **d** £99·68

6. 46p **7.** £175·20

8. a £151·20 **b** £1044 **c** £264 **d** £40·80

9. a £816 **b** £652·80

10. £6·30

Page 73   Task E5.1

1. C **2.** A **3.** 1·05

4. a £117 **b** £152 **c** £56 **d** £288

5. £780 **6.** £101 400 **7.** £93·84

Page 74   Task E5.2

1. a 1·4 **b** £58

2. a 0·9 **b** £620

3. fridge £280, washing machine £390, freezer £240

4. £8500 **5.** 34 cm

6. radio (£55) was most expensive (other items £45, £32 and £28)

7. 400

Page 75   Task M5.14

1. a 0·47 **b** 0·21 **c** 0·8 **d** 0·36 **e** 0·04 **f** 0·07

2. a 59% **b** 23% **c** 3% **d** 30% **e** 20% **f** 18%

3. $\frac{2}{5}$, 0·4, 40%; $\frac{9}{100}$, 0·09, 9%; $\frac{1}{4}$, 0·25, 25%; $\frac{7}{10}$, 0·7, 70%, $\frac{3}{4}$, 0·75, 75%

4. b, c, d, e, g, h are true

5. 0·4 = 40% > 39% so correct

Page 75   Task M5.15

1. £120 **2.** £720 **3.** £14 900 **4.** £18·67

5. £75 **6.** A by £390 **7.** £101·38 **8.** £1147·50

Page 76   Task M5.16/M5.17

1. a 7 : 4 **b** 5 : 2 **2. a** 3 : 2 **b** 3 : 1

4. 2 : 1 **5.** 7 : 3 **6.** 16 **7.** 15

8. a 4 : 5 **b** 3 : 4 **c** 2 : 5 **d** 7 : 3

e 8 : 7 **f** 4 : 3 : 7 **g** 1 : 8 **h** 1 : 12

9. both 2 : 3

10. a 4 : 5 **b** 4 : 5 **c** 3 : 5 **d** 7 : 2 **e** 7 : 2 **f** 20 : 9

11. Yuri correct (both 4 : 3 : 7)

Page 77   Task M5.18

1. £54   2. £49   3. £2075   4. £41·93

5. £193·20   6. 825 g

7. 550 g cheese, 10 toms, 20 pineapple chunks, $1\frac{1}{4}$ cucumbers

8. 400 g   9. 400 g   10. €186

11. 144 g butter, 60 g sugar, 210 g flour

12. $\frac{3}{4}$ kg box

Page 78   Task M5.19

1. a £30 : £120   b £36 : £24   c 14 g : 35 g

   d 60 : 36   e £540 : £90 : £270   f 20 l, 25 l, 30 l

2. Todd 9, Claire 21   3. Carl £10 000, Rachel £2000

4. p 210°, q 60°, r 90°   5. 25 000

6. 12 ml

Page 79   Task E5.3

1. a 28 l b 10 l c 56 yellow, 16 blue

2. 6 : 5   3. £1320   4. S £18 000, J £45 000

5. 99   6 17   7. 100°   8. 252

9. £70   10. a 3   b 36 times

Page 80   Task M7.1

1. a 17, 20   b add 3   c yes

2. a 56, 68   b add 12   c yes

3. a 24, 19   b subtract 5   c yes

4. a 13, 18   b add 1 more than the time before   c no

5. a 25, 16   b subtract 9   c yes

6. a 60, 85   b add 5 more than the time before   c no

7. a −13, −19   b subtract 6   c yes

8. a −10, −16   b subtract 6   c yes

9. a 7, 15, 23, 31, 39   b 61, 54, 47, 40, 33

   c 19, 14, 9, 4, −1

10. 16, 10   11. 19, 33   12. −1, 3   13. 26, 2

14. a 18   b 22   15.
    a 15   b 21

16. 290   17. a 81   b 121

18. 26, 37   19. 24, 35   20. 30, 42   21. 38, 52

22. 1024   23. 31, 63

Page 81   Task M7.2

1. a Times 4   b Subtract 8   c Subtract 2·5

   d Times 3, subtract 1

2. 3, 5, 7, 9, 11

3. a 8, 11, 14, 17, 20   b 3, 7, 11, 15, 19   c 9, 11, 13, 15, 17

4. $3n + 2$

5. a 2   b 5   c $5n - 3$

6. A → R, B → Q, C → S, D → P

7. a $3n + 4$ b $7n + 2$ c $9n - 8$ d $8n - 2$

   e $34 - 4n$ f $23 - 5n$ g $4n + 4$ h $25 - 3n$

Page 82   Task M7.3

1. b 16   c $w = 2n + 8$ d 108

2. b

n	1	2	3
s	8	15	22

   c $s = 7n + 1$ d 281

3. b

n	1	2	3
s	10	18	26

   c $s = 8n + 2$ d 322

Page 83   Task E7.1

1. $a - 4 = b$ 2. $y - 6 = x$ 3. $\frac{x}{4} = y$

4. $8a = b$ 5. $b = a + 10$ 6. $m = \frac{n}{5}$

7. a $n = m - 9$ b $n = \frac{m}{4}$ c $n = 3m$ d $n = m + 10$

8. $y = 8x, \frac{y}{8} = x; 8y = x, y = \frac{x}{8}$

9. b

10. a $n = \frac{m - 5}{3}$ b $n = \frac{m + 1}{7}$ c $n = 4(m + 6)$

    d $n = \frac{m + 7}{8}$ e $n = 3(m - 9)$ f $n = 5(m + 3)$

Page 84   Task M7.4

1. e $(1, -2)$

2. a $y = 1$ b $x = 1$ c $x = -2$ d $y = -3$

Page 85   Task M7.5

1. $(0, 1)(1, 3)(3, 7)$

2. $(0, 4)(1, 3)(2, 2)(3, 1)$

3. $(0, 0)(1, 4)(2, 8)(3, 12)(4, 16)$

4. $(-1, -6)(0, -4)(1, -2)(2, 0)(3, 2)$

5. $(-2, -5)(-1, -2)(0, 1)(1, 4)$

Page 86   Task M7.6

1. 2   **2.** 1   **3.** A $\frac{1}{2}$ B $\frac{1}{3}$ C 1   D 4   E 3

4. a 2   **b** 3   **c** $1\frac{1}{2}$ **d** $1\frac{2}{3}$

5. A -3 B -1 C -2 D 2   E $-\frac{1}{4}$ F $-\frac{1}{2}$

6. a -2 **b** -1 **c** $-\frac{1}{2}$ **d** $-\frac{3}{4}$

7. a -20 **b** -5

Page 87   Task M7.7

1. a and **b** all -2 **c** y intercept in $y = mx + c$

2. b all $+3$ **c** y intercept $= c$ in $y = mx + c$

Page 88   Task M7.8

1. $y = 5x + 1, y = 5x + 4, y = 3 + 5x$

2. $y = 3x + 2, y = 2 + 4x$

3. a i 8   **ii** 4   **b i** 2   **ii** -6 **c i** 1   **ii** 0

d i 1   **ii** -5 **e i** -2 **ii** 4   **f i** $\frac{1}{4}$ **ii** 3

g i -3 **ii** 2   **h i** 7   **ii** -6 **i i** $\frac{1}{4}$ **ii** 5

j i -1 **ii** 6   **k i** -3 **ii** 5   **l i** $\frac{1}{6}$ **ii** 9

4. A, $y = x + 2$; B, $y = \frac{1}{2}x - 2$; C, $y = -3x - 2$

Page 89   Task M7.9

1. a 9   **b** 11   **c** 5   **d** 3   **e** 12

2. $(-3, 11)(-2, 6)(-1, 3)(0, 2)(1, 3)(2, 6)(3, 11)$

3. $(-3, 8)(-2, 3)(-1, 0)(0, -1)(1, 0)(2, 3)(3, 8)$

4. $(-2, 16)(-1, 4)(0, 0)(2, 16)$

Page 89   Task E7.2

1. a

x	-3	-2	-1	0	1	2	3
x^2	9	4	1	0	1	4	9
$+3x$	-9	-6	-3	0	3	6	9
y	0	-2	-2	0	4	10	18

b $y = 6·75$

2. a

x	-3	-2	-1	0	1	2	3
x^2	9	4	1	0	1	4	9
$+2x$	-6	-4	-2	0	2	4	6
-5	-5	-5	-5	-5	-5	-5	-5
y	-2	-5	-6	-5	-2	3	10

b $x = 1·45$

Page 90   Task M7.10

1. a 2000   **b** 3000   **c** 5000   **d** 1500

e 12 noon to 12:30 pm

2. a 29 °C   **b** 11 °C   **c** 2:48 pm   **d** 3 pm

e 1:30 pm   **f** 22 °C

Page 91   Task M7.11

1. a 15 km   **b** 45 mins   **c** 60 km/h   **d** 70 km/h

2. a 110 km   **b** 09·15   **c** 08·45   **d** 20 km/h

e 100 km/h

Page 92   Task M8.1

1. a likely   **b** even chance   **c** certain

d unlikely   **e** even chance

2.

impossible	unlikely	even	likely	certain

e, b d f c

Page 92   Task M8.2

1. a 30   **b** $\frac{72}{180}$ **c** No   **2. a** $\frac{58}{120}$ **b** Yes

3. a 30

b $1 - 0·1125, 2 - 0·1, 3 - 0·15, 4 - 0·1125, 5 - 0·125,$
$6 - 0·1125, 7 - 0·1375, 8 - 0·15$

c Yes

Page 93   Task M8.3

1. a $\frac{1}{15}$ **b** $\frac{2}{15}$ **c** $\frac{1}{5}$ **d** $\frac{2}{15}$

2. a $\frac{4}{13}$ **b** $\frac{11}{13}$ **c** $\frac{6}{13}$

3. a $\frac{3}{10}$ **b** $\frac{2}{3}$

4. a $\frac{1}{6}$ **b** $\frac{3}{8}$ **c** $\frac{11}{24}$

5. a $\frac{1}{5}$ **b** $\frac{4}{5}$ **c** 0   **d** $\frac{1}{2}$

Page 94   Task E8.1

1. 24

2. **a** 20   **b** 20   **c** 40   **d** 40

3. 64   4. 28   5. 4

6. **a** 7   **b** 35   **c** 63

7. **a** $\frac{3}{5}$ **b** 6   **c** 9

8. $\frac{2}{3}$

Page 95   Task M8.4

1. CT, CA, MT, MA

2. **a** (1, 2)(1, 4)(1, 9)(1, 12)(8, 2)(8,4)(8,9)   **b** 12
   (8, 12)(27, 2)(27, 4)(27,9)(27, 12)

3. **a** BBB, GGG, BBG, BGB, GBB, GGB, GBG, BGG   **b** $\frac{3}{8}$

4. AK, AJ, AT, KJ, KT, JT

5. **a**

×	1	2	3	4
1	1	2	3	4
2	2	4	6	8
3	3	6	9	12
4	4	8	12	16

   b $\frac{3}{16}$

Page 96   Task M8.5

1. **a** 60; 37, 23; 20, 17, 9, 14   **b** $\frac{17}{37}$ **c** $\frac{9}{23}$

2. **a** 96; 58, 38; 46, 12, 25, 13   **b** $\frac{38}{96} = \frac{19}{48}$ **c** $\frac{13}{38}$

3. **a** 800; 460, 340; 60, 400, 275, 65   **b** $\frac{400}{460} = \frac{20}{23}$

4. **a** 300; 165, 135; 63, 102, 27, 108   **b** $\frac{102}{165} = \frac{34}{55}$

Page 97   Task M8.6

1. 0·8   2. $\frac{3}{4}$ 3. **a** 0·7   **b** 0·2   4. $\frac{10}{13}$

5. 0·85   6. **a** 0·15   **b** 0·4   **c** 8

Page 98   Task M8.7

1. **a** 132   **b** 207   **c** 28

2. **a** 35   **b** 53   **c** 14   **d** 22

3. **a** 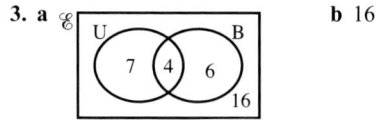   **b** 16

4. **a** 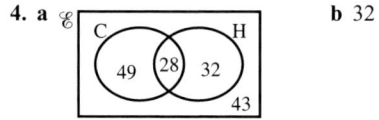   **b** 32

Page 99   Task M8.8

1. **a** $\frac{57}{117} = \frac{19}{39}$ **b** $\frac{32}{117}$ **c** $\frac{28}{117}$

2. **a** $\frac{9}{22}$ **b** $\frac{3}{22}$ **c** $\frac{4}{22} = \frac{2}{11}$

3. **a** $\frac{8}{38} = \frac{4}{19}$ **b** $\frac{19}{38} = \frac{1}{2}$ **c** $\frac{25}{38}$

4. **a** $\frac{18}{85}$ **b** $\frac{16}{85}$

5. **a** $\frac{85}{185} = \frac{17}{37}$ **b** $\frac{148}{185} = \frac{4}{5}$ **c** $\frac{25}{185} = \frac{5}{37}$

Page 100   Task M9.1

1. 1·30   2. 6·15   3. 2·20

4. 6·45   5. 4·55   6. 3·35

7. **a** 14   **b** 22   **c** 36   **d** 48   **e** 52
   f 26   **g** 38

8. **a** 580   **b** 630   **c** 760   **d** 54 *l*   **e** 160 g

Page 101   Task M9.2

1. **a** 1·2   **b** 2·4   **c** 3·6   **d** 4·4
   e 5·8   **f** 3·2   **g** 3·4

2. **a** 1·75 kg   **b** 0·18 kg   **c** 3·4 *l*   **d** 3·20 *ml*

3. **a** 6·25 kg   **b** 7·75 kg   **c** 15·4 *l*   **d** 16·6 *l*
   e 0·018 kg   **f** 0·032 kg

4. 8 litres

Page 102   Task M9.3

1. **a** 300   **b** 120   **c** 580   **d** 364   **e** 70

2. **a** 4000   **b** 6500   **c** 3200   **d** 4718   **e** 900

3. **a** metres   **b** tonnes   **c** millimetres

4. No

5. **a** 6   **b** 4·5   **c** 5000   **d** 5300   **e** 7186

6. **a** 8000   **b** 9500   **c** 4600   **d** 4315   **e** 60

7. 440 *ml*   8. 2100 m

Page 103   Task M9.4

1. 1680 g   2. 2·863 km

2. **a** 260 cm   **b** 382 cm   **c** 4·7 m   **d** 9 cm

   e 0·4 cm   **f** 1·5 km   **g** 3500 g   **h** 0·6 kg

   i 280 g   **j** 1900 kg   **k** 0·62 *l*   **l** 1 937 000

   m 8200 *ml*   **n** 3260 *ml*   **o** 0·043

4. 80

5. **a** 7·4 cm, 8 cm, 83 mm, 0·81 m

   b 0·7 kg, 738 g, 780 g, 0·79 kg

   c 57 m, 509 m, 0·6 km, 4·7 km, 5 km

   d 274 *ml*, 275 *ml*, 0·279 *l*, 0·28 *l*, 2·14 *l*

Page 103   Task M9.5

1. **a** T   **b** F   **c** T   **d** T   **e** F   **f** T

2. 10·05 pm   3. 10:50   4. 60

5. 1 hr 45 mins   6. yes

7.

	Bus 1	Bus 2	Bus 3	Bus 4	Bus 5
Bus Station	07:50	08:35	08:55	09:45	10:25
Cinema	07:59	08:44	09:04	09:54	10:34
Town Hall	08:10	08:55	09:15	10:05	10:45
Cherry Hill	08:22	09:07	09:27	10:17	10:57
Train Station	08:35	09:20	09:40	10:30	11:10

Page 104   Task M9.6

1. **a** 20 mph   **b** 65 km/h

2. **a** 90 miles   **b** 29 miles

3. **a** 4 hrs.   **b** $2\frac{1}{2}$ hrs.

5.

Dist (km)	Time (hrs)	Speed (km/hr)
324	9	36
204	4	51
90	1·5	60
150	6	25
245	7	35
40	0·5	80

6. 12 km/h   2. 126 km

Page 105   Task E9.1

1. 52 mph, 42 mph, 48 mph, 90 mph, 85 mph

2. 141 km/h   3. 8:50 am

4. Both travel the same distance

5. **a** 5 m/s   **b** 20 m/s   **c** 54 km/h   **d** 126 km/h

6. 36 km/h   7. 1 hour 30 minutes   8. 900 m

Page 106   Task M9.7

1. **a** 8 cm   **b** 9 cm   **c** 10 cm

2. Various. For example: 3 × 4, 2 × 5 etc.

3. 6 cm

4. **a** *a* 10 cm, *b* 8 cm   **b** 58 cm

5. 44 cm   6. 36 cm   7. 50 cm

8. Should be cm not cm²

9. 9 cm   10. 6 cm   11. 4 cm

Page 107   Task M9.8

1. 24 cm²   2. 6 cm   3. 52 cm²   4. 96 cm²

5. 102 cm²   6. 84 cm²   7. 180 cm²   8. 320 cm²

9. Change 2 m to 200 cm

10. £183·20

Page 109   Task M9.9

1. 60 cm²   2. 125 cm²   3. 72 cm²   4. 16 cm

5. 12 cm   6. 1 : 4   7. $\frac{4}{7}$ m²   8. 460 cm²

Page 110   Task M9.10

1. 23 cm   2. 38 mm

3. **a** 37·7 cm   **b** 18·8 m   **c** 56·5 cm   **d** 53·4 m

4. No. 75 cm less than circumference 78·5 cm

5. Triangle   6. A

Page 111   Task E9.2

1. 41·1 cm   2. 79·7 cm   3. 17·9 cm   4. 38·8 cm

5. 300·5 cm   6. 16·7 m   7. 50   8. 1515 times

9. 28

Page 112   Task M9.10

1. $78.5\,cm^2$ 2. $254.5\,cm^2$ 3. $113.1\,cm^2$ 4. $804.2\,cm^2$

5. **a** radius   **b** chord   **c** diameter   **d** radius

6. $380.1\,m^2$ 7. the circle   8. $13.7\,cm^2$

9. **a** 81π **b** 16π **c** 49π

Page 113   Task E9.3

1. $226.2\,cm^2$ 2. $22.9\,cm^2$ 3. $1289.1\,cm^2$ 4.
$3.14\,cm^2$

5. $103.7\,cm^2$ 6. $84.1\,cm^2$ 7. $50.4\,cm^2$ 8. $102.5\,cm^2$

9. $18\pi\,cm^2$

Page 114   Task M9.11

1. Front = $36\,cm^2$, back = $36\,cm^2$, top = $54\,cm^2$,
   bottom = $54\,cm^2$, side 1 = $24\,cm^2$, side 2 = $24\,cm^2$;
   total = $228\,cm^2$

2. $216\,cm^3$

3. **a** A   **b** $42\,cm^2$ **c** B   **d** $70\,cm^3$

4. 2 : 5   5. cm^3 not cm

6. $1\,050\,000\,cm^3$ or $1.05\,m^3$

Page 115   Task M9.12

1. $270\,cm^3$ 2. $360\,m^3$ 3. $525\,cm^2$

4. $800\,cm^3$ 5. $4170\,cm^3$ 6. $37\,cm^2$

7. Declan correct   8. $660\,cm^2$

Page 116   Task M9.13

1. $1570.8\,cm^3$ 2. $2290.2\,cm^3$ 3. $364.4\,cm^3$

4. $1055.6\,cm^3$ 5. 240π 6. 7.5 litres

7. **a** 982 secs.   **b** 16 minutes

8. 10

Page 118   Task M9.14

2. Q, S

3. All sides in the same ratio in both shapes and angles are
   equal.

4. 42 cm

5. All sides not in same ratio.

6. **a** 12.5 cm   **b** 6 cm   **c** 8 cm   **d** 17.5 cm

Page 119   Task M10.1

1. 2, 4, 6, 7, 12, 14, 14, 18, 19 → Median = 12

2. **a** 3   **b** 2, 6

3. 80

4. **a** 8   **b** 9   **c** 10   **d** 14

5. **a** 9   **b** 8   **c** 4, 8   **d** 12

6. **a** 6   **b** 6.5   **c** 9   **d** 7

7. The Saxo

8. 1.8   9. 25

Page 120   Task E10.1

1. **a** 48   **b** 40
   c median: 40 better represents 6 of the 7 marks

2. **a** 9   **b** 9.1   **c** mode: 7 out of 10 scores were 9

3. 1, 4, 7, 9, 14   4. 50%   5. 222

6. **a** 693 kg   **b** 61 kg

7. 10

8. Median. It is the salary of a person in the middle of the
   range of salaries.

9. 3, 3, 8, 9, 12 or 3, 3, 8, 10, 11

10. £27 200

Page 121   Task M10.2

1. **a** 60   **b** 45   **c** 70   **d** 2013
   e 2011, 2014   **f** 28   **g** 30
   h Hotley Albion by 7 goals

2. **b** $\frac{1}{3}$

3. **a** 26   **b** 33   **c** 38   **d** 2005
   e 5   **f** 16

Page 122   Task M10.3

1. **a** £32   **b** £37

2. **b** 38 kg   **c** $\frac{3}{11}$ **d** 31.8%

3. **a** 174 cm, 36 cm   **b** 175 cm, 33 cm
   c Tampton Trojans are slightly taller with a smaller
   spread of heights.

Page 123   Task M10.4

1. **a** 90   **b** 4°

   c bus 120°, car 40°, tube 100°, foot 60°, bike 40°

2. Adventure 48°, comedy 144°, horror 56°, romance 16°, cartoon 96°

3. Blue 69°, green 24°, red 84°, yellow 123°, purple 12°, other 48°

Page 124   Task M10.5

1. **a** 100   **b** 75   **c** 125

2. **a** 144°   **b** 54°   **c** 126°

3. **a** 40   **b** 24   **c** 96

4. We do not know the number of students at each school.

Page 125   Task M10.6

1. **a**

	Drive	Not Drive	Total
Female	110	60	170
Male	180	50	230
Total	290	110	400

b 50

2.

	E	M	S	Total
Boys	86	73	89	248
Girls	220	58	74	352
Total	306	131	163	600

220 girls chose English.

3. **a**

	Film on DVD	Cinema	Theatre	Total
Male	1	5	2	8
Female	6	4	2	12
Total	7	9	4	20

b 25%

4.

	Stay	College	Leave	Total
Year 10	86	128	26	240
Year 11	120	109	31	260
Total	206	237	57	500

a 48%   **b** 47·4%

Page 126   Task M10.7

1. Negative correlation

2. **a** No correlation

3. **b** No correlation   **c** $\frac{1}{4}$

Page 127   Task M10.8

1. What was being sold? Did they work the same amount of time?

2. e.g. large intervals on the vertical axis make the differences between each bar appear smaller than they really are.

3. The area of the larger triangle is more than 1·25 times the area of the smaller triangle.

4. Labelling on vertical axis not consistent, month 3 is missing, month 8 is not included in the line graph.

5. Vertical scale does not start at 0, gaps between bars not equal, no label on vertical axis, bars are not the same width, heading not complete.

6. Implies it is the best chocolate in the world.

Page 128   Task M10.9

1. **a** 61 kg   **b** 61 kg

2. **a** 5   **b** 4

3. **a** $14\frac{1}{2}$ to $15\frac{1}{2}$ **b** 16 to $16\frac{1}{2}$

4. **a** Chetley Park 2 to 5, Wetton 6 to 9

   b Wetton: higher median number of visits.

Page 129   Task M10.10

1. **a** 82   **b** 2·05

2. **a** 180   **b** 1·7

3. **a** 2·64   **b** 2

Page 130   Task M10.11

1. **a** 415   **b** 5

   c Assume that the number of meals is at the middle of each group.

2. **a** Batton City   **b** 43·8   **c** 4

Page 131   Task M10.12/M10.13

1. Baker median £5·75, range £1·95
   Butcher median £5·95, range £2·65

2. **a** mean 3·32 range 7   **b** mean 3·13 range 7

3. 8C mode 3, range 6; 8D mode 2, range 7

4. Car A: mean = £6625, range = £4000
   Car B: mean = £7350, range = £5400

5. **a** 18·79   **b** 11

   c The 8 year-olds received more presents.

Page 133   Task M11.1

1. **a** 5·4 cm   **b** 6·3 cm   **c** 1·8 cm
   d 8·3 cm   **e** 14·5 cm   **f** 14·6 cm

2. **b** is larger by 1·6 cm

3. **a** is smaller by 1·1 cm

Page 134   Task M11.2

1. 57°   2. 56°   3. 40°
4. 120°   5. 33°   6. 125°
7. 55°   8. 115°/116°   9. 30°
10. **a** ac   **b** ac   **c** ac   **d** ac
    e obt   **f** obt   **g** obt   **h** ac

Page 134   Task M11.3

2. 6·5/6·6 cm   3. 80°   4. 6·2 cm
5. 4·2 cm   6. 8·7 cm

Page 135   Task M11.4

1. 35°   2. 44°   3. 66°/67°   4. 50°/51°   5. 59°

6. eg. 6 cm, 5 cm or 10 cm, 3 cm

8. **a** $x = 38°, y = 6·7$ **b** $x = 81°, y = 60°$

Page 136   Task M11.5

4. **a** 23·1 m × 11·4 m   **b** 8·7 m (or 17·4 m)
   c 5·1 cm   **d** 75·7 m²

5. **a** 53 km   **b** 42 km   **c** 37 km

Page 138   Task M11.6

2.   3. 12 cm³

4. 144

5. B by 2 cm³

6. 6

Page 139   Task E11.1

1. A and C make tetrahedrons

Page 141   Task M11.7

1. **a** 10   **b** 9

3. 6

Page 142   Task M11.8

1. triangular prism. 5 faces.

2. 30 cm²

3. square based pyramid

4. 24 cm³

5. 12 cm³

6. 192 cm²

Page 143   Task M11.9

1. 8 cm   2. 2 km

3. 7 cm → 4·2 m, 5 cm → 100 m, 8 cm → 4 km,
   3 cm → 3 km, 8 cm → 320 m,   2·5 cm → 125 km

4. 5 cm

5. 12 m

6. **a** 6·4 km   **b** 5 km   **c** 4·4 km

7. 50 km²

Page 144   Task M11.10

1. A 035°, B 105°, C 220°, D 340°

2. **a** 060°   **b** 240°   **c** 180°   **d** 300°   **e** 120°

3. **a** 115°   **b** 150°   **c** 220°   **d** 295°   **e** 330°

Page 145   Task M11.11 _____

1. **a** 236°   **b** 056°
2. **a** 065°   **b** 130°   **c** 165°   **d** 310°   **e** 245°   **f** 345°
3. **a** 5·4 km
4. **b** 92 km
5. **b** 245°   **c** 5 km

Page 147   Task M11.12 _____

1. **a** (3, 3)   **c** 12
2. $\left(3, 4\frac{1}{2}\right)$
3. **a** (4, 4)   **b** (−4, 1)   **c** (−1, −4)
4. **a** (5, 3), (3, 5), (−2, 3), (3, 1)   **b** 14   **c** 16
5. **c** (5, 2)   **d** (2, 0)

ESSENTIAL GCSE MATHS 4–5 ANSWERS

Page 1 Task M1.1 _____

1. 7, 13

2. a 2 **b** 5 **c** 3 **d** 13 **e** $\frac{1}{2}$ **f** 1

3. £26

4. £5

5. a 0·24 **b** 0·035 **c** 4·97 **d** 1·2

e 0·09 **f** 27 **g** 72 **h** 14·81

6. 44

7. £1424·88

8. 7

Page 1 Task M1.2 _____

1. a 4 **b** 2 **c** $\frac{1}{5}$ **d** 3·75

2. A → R, B → U, C → S, D → P, E → Q, F → T

3. $112·90

4. A

5. 1^9, 3^4, 5^3, 2^7, 4^4

6. a 4·14 **b** 3·49 **c** 4·91

d 4·64 **e** −0·894 **f** 0·0206

7. ≈0·5 × 12 = 6 so probably correct

8. $2\frac{3}{10}$

9. $3\frac{1}{9}$ cm^2

10. a ≈300 **b** ≈4 **c** ≈2

Page 3 Task M1.3 _____

1. a F **b** F **c** F **d** T **e** T **f** F

2. a 6^3 **b** 9^{12} **c** 7^3 **d** 5^7 **e** 2^5 **f** 8^5

3. 3^5 mm

4. a 3^2 **b** 7^5 **c** 6^2 **d** 2^{10} **e** 5^7 **f** 3^5

5. 2^9

6. Q

7. a 72 **b** 2000 **c** 13 824

8. 2^{13} cm^2

Page 4 Task E1.1 _____

1. a $\frac{1}{4}$ **b** $\frac{1}{8}$ **c** $\frac{1}{36}$ **d** $\frac{1}{64}$

e $\frac{1}{100}$ **f** $\frac{1}{125}$ **g** $\frac{1}{7}$ **h** $\frac{1}{64}$

2. a 4^{-2} **b** 3^{-4} **c** 2^{-5} **d** 8^{-3}

3. 5^{-2}

4. 3^{-3}

5. a T **b** T **c** F **d** F **e** T **f** T

6. $\frac{9}{20}$

7. Reciprocal of 2 greater by $\frac{1}{6}$

8. a 3 **b** $\frac{7}{3}$ **c** $\frac{25}{9}$ **d** $\frac{49}{4}$

9. 9^{-2}

10. a 5 **b** 32 **c** 10

11. 10^{-1}

12. No, e.g. reciprocal of 0·5 is 2

Page 5 Task M1.4 _____

1. a 11 **b** 20 **c** 3 **d** 3 **e** 9 **f** 16

2. 8·5 cm^2

3. 9

4. a 4 **b** 3 **c** 7 **d** 5

5. B greater by 2

6. 59 cm^2

7. a 3·0 **b** 4·0 **c** 0·3 **d** 2·7 **e** 0·7 **f** 10·6

8. 7·1 cm^2

Page 6 Task M1.5 _____

1. 350 = 2 × 5^2 × 7, 396 = 2^2 × 3^2 × 11,
440 = 2^3 × 5 × 11, 405 = 3^4 × 5

2. a 2^5 × 3 **b** 2 × 3 × 5^2 **c** 2 × 5 × 31 **d** 2^3 × 5 × 13

4. 841

5. e.g. 64

6. 525

Page 7 Task M1.6 _____

1. a 18 **b** 6552

2. a 20 **b** 2100

3. a HCF = 28, LCM = 4620

   b HCF = 35, LCM = 55 125

   c HCF = 99, LCM = 103 950

4. a $2\,mn^2$ **b** $12\,m^3n^3$

5. 16 and 24

6. $60\,a^2b^3c$

7. $x = 7,\ y = 5$

8. 8 loaves of bread and 11 packs of ham slices

Page 8   Task M1.7

1. a $4{\cdot}7 \times 10^4$ **b** $7{\cdot}6 \times 10^3$ **c** $5{\cdot}75 \times 10^2$

   d $7{\cdot}4 \times 10^5$ **e** $3{\cdot}69 \times 10^5$ **f** $3{\cdot}2 \times 10^{-3}$

   g $8{\cdot}9 \times 10^{-2}$ **h** $8{\cdot}64 \times 10^{-1}$

2. $3{\cdot}8 \times 10^4$

3. a 500 000   **b** 6000   **c** 0·04   **d** 18 000

   e 830 000   **f** 0·0057   **g** 2380   **h** 0·429

4. $2{\cdot}9 \times 10^4$

5. $7{\cdot}68 \times 10^{-4},\ 7{\cdot}4 \times 10^{-3},\ 7{\cdot}42 \times 10^{-3}$

6. a 3×10^{-2} **b** $3{\cdot}08 \times 10^2$ **c** 5×10^6

   d 4×10^{-4} **e** $2{\cdot}287 \times 10^2$ **f** 2×10^8

   g $4{\cdot}673 \times 10$ **h** $9{\cdot}2 \times 10^{-2}$ **i** $7{\cdot}8 \times 10^7$

   j $9{\cdot}6 \times 10^{-1}$ **k** 3×10^{-2} **l** $6{\cdot}2 \times 10^5$

7. £286 400 or £($2{\cdot}864 \times 10^5$)

Page 9   Task M1.8

1. a $7{\cdot}3 \times 10^5$ **b** $4{\cdot}2 \times 10^{15}$ **c** 8×10^6

   d $3{\cdot}2 \times 10^{23}$ **e** $6{\cdot}8 \times 10^{-5}$ **f** $3{\cdot}74 \times 10^{-5}$

   g $4{\cdot}25 \times 10^{40}$ **h** $5{\cdot}6 \times 10^{-8}$

2. 8×10^9

3. a 5×10^{15} **b** 6×10^9 **c** 7×10^5

   d $6{\cdot}8 \times 10^{-26}$ **e** $1{\cdot}2 \times 10^{20}$ **f** $3{\cdot}6 \times 10^{46}$

   g 2×10^{15} **h** $3{\cdot}5 \times 10^{12}$ **i** 2×10^{37}

   j $1{\cdot}6 \times 10^{11}$ **k** $2{\cdot}9 \times 10^{28}$ **l** 5×10^{34}

4. $4{\cdot}3 \times 10^5$

5. a $5{\cdot}7 \times 10^6$ **b** $8{\cdot}6 \times 10^8$ **c** $4{\cdot}61 \times 10^{12}$

   d $2{\cdot}1 \times 10^{-2}$ **e** $2{\cdot}43 \times 10^{-7}$ **f** $6{\cdot}96 \times 10^{24}$

6. $6 \times 10^{26}\,\text{cm}^3$

7. $4{\cdot}14 \times 10^4$

8. a Rosa Williams   **b** $4{\cdot}11 \times 10^6$

9. $3{\cdot}43 \times 10^{-34}$

10. $3{\cdot}62 \times 10^5$

Page 10   Task M1.9

1. £($2{\cdot}518 \times 10^7$)

2. a $3{\cdot}02 \times 10^{36}$ **b** $6{\cdot}94 \times 10^{-57}$

   c $1{\cdot}57 \times 10^{40}$ **d** $1{\cdot}12 \times 10^{24}$

   e $3{\cdot}59 \times 10^{13}$ **f** $2{\cdot}28 \times 10^{24}$

   g $4{\cdot}86 \times 10^{66}$ **h** $4{\cdot}09 \times 10^{14}$

3. $1{\cdot}058 \times 10^{10}$

4. $4{\cdot}1 \times 10^7\,\text{m}$

5. $1{\cdot}68 \times 10$

6. a F   **b** D

   c E, A, F, B, D, C   **d** $2{\cdot}73 \times 10^6$

7. $7{\cdot}9 \times 10^{-3}$

8. 53·8%

9. $1{\cdot}01 \times 10^9$

Page 11   Task M2.1

1. $18mn$

2. $7n$

3. $12a^2$

4. m

5. a^2

6. $54mn$

7. Not like terms

8. P is correct

9. a $4m + 28$ **b** $15a + 25$ **c** $2n - 14m$

   d $n^2 - 3mn$ **e** $2m^2 + 9m$ **f** $6a^2 - 3a$

   g $5m^2 - 10mn$ **h** $12a^2 + 30ab$

10. a $4m + 11n$ **b** $3n + 2m + 1$ **c** $5n + 3m + 2n^2$

11. $13a + 5b$

12. $36m + 26$

Page 12   Task M2.2

1. 7   **2.** 72   **3.** 8

4. 42   **5.** 5   **6.** 4

7. Expand $7(3m + 5n) = 21m + 35n$

8. $9(2a - 3b)$ **9.** $4(3m + 4n)$ **10.** $m(m + 3)$

11. $a(a - 7b)$ **12.** $x(2y - x)$ **13.** $2n(4n + 3m)$

14. $3^9\,\text{cm}^2$ **15. a** $3n + 135 = 180$ **b** $n = 15°$

16. a $9n + 90 = 180$ **b** $n = 10°$

17. 5^6 **18.** 6^3 **19.** 2^3

20. 7^4 **21.** 3^3 **22.** 8^3

23. $34n + 42$ **24.** $38x + 48$

25. Pat. Both equal $72mnp$

Page 13   Task E2.1 ——————————

1. $13n + 19$

2. a m^5 **b** n^{10} **c** $28n^9$ **d** $12m^3$ **e** n^3 **f** $4m^2$

3. a $5a - 20 = 60$ **b** $a = 16$

4. a F   **b** F   **c** T   **d** T   **e** F   **f** T

5. Correct

6. a $-\dfrac{1}{3}$ **b** $-\dfrac{2}{7}$ **c** -4 **d** -2 **e** 4 **f** $-\dfrac{1}{2}$

7. $40a + 20$

8. S

9. a $2n(2n - 5m)$ **b** $15p(2q - p)$ **c** $4m(4m + 3)$
     d $4x(9x + 5y)$ **e** $5m(2 + n + 4m)$ **f** $8y(7x - 5y)$

10. $6n$

Page 15   Task E2.2 ——————————

1. $3a + 19$ **2.** $22b + 36$ **3.** $25a + 42$

4. $28y + 8$ **5.** $17n + 27$ **6.** $31b + 30$

7. $34c + 25$ **8.** $130x + 70$ **9.** $10a + 4$

10. $13x + 51$ **11.** $2a + 14$ **12.** $7m + 3$

13. $12y + 18$ **14.** 42 **15.** $2a + 21$

16. $3x + 1$ **17.** $26n + 43$ **18.** $32q + 31$

19. $11x^2 - x$ **20.** $7n^3 - 30n$

21. $8m^2 - 3n^2 - 16mn$

22. $12a^2 - 14ab - 9ac + 8bc$

Page 16   Task M2.3 ——————————

1. $C = (110x + 48)p$

2. $P = 11x + 8$

3. $A = 2m - 15$

4. $A = 2x^2$

5. $T = 120n + 80$

6. $C = 45x + 0{\cdot}25y$

7. $P = 9n - 200$

8. $P = 16n + 40$

9. a $3n + 20$ **b** $T = 7n + 20$

10. $V = 16\,640n^2$

Page 17   Task M2.4/E2.3 ——————————

1. a 46 **b** 150

2. a 108 **b** 300 **c** 768

3. $1{\cdot}35 \times 10^{18}$

4. a 65 **b** 7 **c** 23 **d** 20

5. a 497 **b** -136

6. a 114 **b** $20{\cdot}2$

7. $16{\cdot}1$

8. a 100 **b** 50 **c** $6{\cdot}25$

Page 18   Task M2.5 ——————————

1. a $x^2 + 7x + 10$ **b** $x^2 + 10x + 21$
   c $x^2 + 10x + 16$

2. $x^2 + 10x + 24$ **3.** $m^2 + 12m + 35$

4. $y^2 + 14y + 40$ **5.** $n^2 + 13n + 42$

6. $a^2 + 11a + 18$ **7.** $x^2 + 15x + 36$

8. $x^2 + 8x + 16$ **9.** $x^2 + 12x + 36$

10. $x^2 + 14x + 49$ **11.** $x^2 + 2x + 1$

12. $x^2 + 18x + 81$ **13.** $y^2 + 20y + 100$

14. $m^2 + 10m + 25$

Page 19   Task E2.4 ——————————

1. a $x^2 - 2x - 15$ **b** $y^2 - 8y + 16$
   c $n^2 - 3n - 40$

2. $x^2 - 2x - 24$ **3.** $a^2 - 11a + 30$

4. $y^2 - 10y + 21$ **5.** $n^2 - 5n - 36$

6. $m^2 + m - 42$ **7.** $b^2 - 3b - 40$

8. $a^2 - 16a + 64$ **9.** $x^2 - 7x + 12$

10. $f^2 + 3f - 70$ **11.** $n^2 - 10n + 25$

12. $y^2 - 14y + 49$ **13.** $a^2 - 4a + 4$

14. $x^2 + 11x + 28$ **15.** $p^2 - 6p + 9$

16. $-m^2 + 6m + 16$ **17.** Both areas $= x^2 + 16x + 64$

18. $15x^2 + 22x + 8$ **19.** $10a^2 + 13a + 4$

20. $6n^2 + 2n - 28$ **21.** $21y^2 - 32y + 12$

22. $16a^2 + 48a + 36$ **23.** $25m^2 - 90m + 81$

24. $5y^2 + 36y + 36$ **25.** $63 - 10c - 8c^2$

26. $32x^2 - 4xy - 6y^2$ **27.** $2m^2 + 22m + 65$

Page 20   Task M2.6

1. 2

2. Dave is correct because $n = 4$

3. 2

4. All values of x

5. **a** B, E   **b** D   **c** A, G
   d F, H   **e** C

6. **a**, **c** and **d** are correct

7. **a** identity   **b** identity   **c** equation ($x = 6$)
   d identity

8. True for all values of x so an identity

Page 21   Task M2.7

1. $4n = 12, n = 3$ 2. $6n + 9 = 15, 6n = 6, n = 1$

3. $4n - 6 = 14, 4n = 20, n = 5$

4. 4   5. 2   6. 11   7. 9

8. 8   9. 2   10. 3   11. 1

12. 10   13. 3   14. 5   15. 4

16. 5   17. 4   18. 7   19. 6

Page 22   Task E2.5

1. $2n + 10 = 11, 2n = 1, n = \frac{1}{2}$

2. $6n + 9 = 3, 6n = -6, n = -1$

3. $5x + 10 = 8, 5x = -2, x = -\frac{2}{5}$

4. $-\frac{3}{2}$ 5. $-\frac{4}{5}$ 6. $\frac{5}{6}$ 7. 0·3

8. $\frac{19}{4}$ 9. -2 10. -3 11. -1

12. $-\frac{1}{6}$

Page 22   Task M2.8

1. 7   2. 3   3. 5   4. 4

5. 9   6. -1 7. 3   8. 6

9. 5   10. 7   11. 7   12. 5

13. 16   14. 8   15. 2   16. 10

17. 9   18. 6   19. 9   20. 7

Page 23   Task E2.6

1. $3x = -2, x = -\frac{2}{3}$

2. $6x + 12 = 4x + 24, 2x = 12, x = 6$

3. $2n = -6, n = -3$

4. $\frac{1}{3}$ 5. $\frac{4}{5}$ 6. 4   7. 7

8. -2 9. -1 10. 10   11. 3

12. 20   13. 12   14. 9   15. -7

16. 2   17. 3   18. -11

Page 24   Task M2.9

1. **a** $6x = 180°$ **b** $30°$ **c** $30°, 60°, 90°$

2. **a** $9x + 6 = 60$ **b** 6 cm

3. 8

4. **a** $4x + 14 = 50$ **b** 9   **c** 16 cm $\times$ 9 cm

5. £18   6. 7 cm   7. 6 cm

8. 159 cm

Page 25   Task E2.7

1. 24 cm^2

2. **a** $20x + 220 = 360$ **b** 7   **c** $81°, 105°, 79°, 95°$

3. **a** $n + 1, n + 2$ **b** $3n + 3 = 144$ **c** $n = 47, 48, 49$

4. **a** $x = 12$ **b** 135 cm^2

5. 17, 18, 19, 20

6. **a** 4   **b** 37 cm

7. 34 m/s

8. 64 cm $\times$ 165 cm

Page 26   Task E2.8

1. 12   2. 10   3. 30

4. 56   5. 27   6. 48

7. 19   8. 28   9. 8

10. Should be $\frac{6}{12} = \frac{1}{2}$. Using $x = 2$ in given equation gives
$\frac{6}{2} - 8 = -5$ which is wrong.

11. 8   12. 6   13. 7

14. 4·5   15. 21   16. $\frac{5}{7}$

17. 8   18. $\frac{4}{5}$ 19. -36

Page 27   Task M2.10

1. $(x + 5)(x + 7)$ 2. $(m + 3)(m + 9)$

3. $(y - 3)(y - 1)$ 4 $(n - 6)(n + 4)$

5. $(a - 9)(a + 3)$

6. $(c - 10)(c + 2)$

7. $(n - 8)(n - 3)$

8. $(y - 9)(y - 5)$

9. $(a + 6)(a - 5)$

10. $(x + 8)(x - 9)$

11. $(p + 11)(p + 4)$

12. $(m + 10)(m - 6)$

13. $(a - 8)(a - 7)$

14. $(q + 12)(q - 8)$

15. $(b - 15)(b + 10)$

16. $(x + 1)^2$, $a = 1$

17. $(x + 4)^2$, $b = 4$

18. $(x - 3)^2$, $c = 3$

19. $(x - 7)^2$, $d = 7$

20. $(x + 3)(x + 2)$

21. $(x - 6)(x + 2)$

22. Incorrect. AB $= x + 4$.

3. a $3\frac{1}{5}$ **b** $3\frac{6}{7}$ **c** $3\frac{3}{4}$

d $10\frac{3}{10}$ **e** $6\frac{5}{8}$

4. $\frac{3}{25}$, 0.16, $\frac{1}{5}$, 0.25

5. $\frac{1}{6}$

6. a $0.86, 0.858, 0.839, 0.089, 0.087$

b $0.86, 0.858, 0.839, \frac{4}{5}, 0.089, 0.087$

7. Two are true. Not true are $0.65 > \frac{13}{20}$, $0.65 \neq \frac{13}{20}$, $0.65 < \frac{13}{20}$

8. 0.203

Page 27   Task M2.11

1. $(x + y)(x - y)$

2. $(b + 3)(b - 3)$

3. $(y + 5)(y - 5)$

4 $(a + 8)(a - 8)$

5. $(n + 2)(n - 2)$

6. $(p + 1)(p - 1)$

7. $(6 + x)(6 - x)$

8. $(3y + z)(3y - z)$

9. $(7 + 2a)(7 - 2a)$

10. $(7x + 9y)(7x - 9y)$

11. $(12m + 5)(12m - 5)$

12. $\left(4b + \frac{1}{3}\right)\left(4b - \frac{1}{3}\right)$

13. $5(x + 2)(x - 2)$

14. $3(2a + 3b)(2a - 3b)$

15. $4(m - 5)(m + 3)$

16. $3(n + 4)(n - 4)$

17. $2(5 + b)(5 - b)$

18. $5(t + 1)(t + 2)$

19. $6(n - 5)(n - 2)$

20. $3(2p + 7)(2p - 7)$

21. $4(x - 6)(x + 2)$

22. $(3\pi + 1)(3\pi - 1)$

23. $\left(\frac{1}{2}e + 5\right)\left(\frac{1}{2}e - 5\right)$

24. $(e + 4\pi)(e - 4\pi)$

25. 228

26. $(x + 3)(x - 3)$

27. $16x$

Page 28   Task M3.1

1. 20 mins

2. $\frac{4}{5} = \frac{24}{30}$, $\frac{5}{6} = \frac{25}{30}$ so correct. $\frac{4}{5} < \frac{5}{6}$

3. 6

4. $\frac{11}{16}$

5. 0.15

6. $\frac{28}{30}$

7. $\frac{3}{25}, \frac{3}{20}, \frac{45}{200}, \frac{1}{4}, \frac{3}{10}, \frac{5}{16}$

8. Plain flour 280 g, butter 150 g, sugar 525 g

Page 29   Task M3.2

1. correct

2. A

Page 30   Task M3.3

1. a $\frac{7}{10}$ **b** $\frac{7}{20}$ **c** $\frac{7}{200}$ **d** $\frac{23}{25}$

e $\frac{309}{500}$ **f** $\frac{637}{2000}$ **g** $\frac{713}{1000}$ **h** $\frac{5}{8}$

2. correct

3. a $0.\dot{2}$ **b** $0.41\dot{6}$ **c** $0.8\dot{3}$ **d** $0.\dot{3}8461\dot{5}$

e $0.\dot{8}5714\dot{2}$

4. $0.427, 0.428, \frac{3}{7}$ **5.** $1.003\dot{6}$

Page 31   Task M3.4

1. a $\frac{11}{12}$ **b** $\frac{11}{24}$ **c** $\frac{11}{40}$ **d** $\frac{41}{63}$

2. $\frac{3}{40}$ mile

3. a $4\frac{1}{6}$ **b** $5\frac{1}{6}$ **c** $4\frac{11}{24}$ **d** $1\frac{5}{12}$

4. $1\frac{23}{24}$ km   **5.** $1\frac{2}{3}$ litres   **6.** $\frac{33}{56}$

Page 31   Task M3.5

1. a $\frac{5}{9}$ **b** $-\frac{1}{8}$ **c** 15 **d** $\frac{2}{3}$

e $1\frac{1}{3}$ **f** $1\frac{3}{4}$ **g** 6 **h** $-4\frac{3}{8}$

2. $\frac{4}{7}$ **3.** $\frac{3}{8}$ **5.** $96 \, \text{cm}^2$

6. a $\frac{5}{7}$ **b** $1\frac{1}{6}$ **c** $2\frac{2}{3}$

7. $1\frac{23}{40}$ m   **8.** $90 \, \text{m}^2$

Page 33   Task M4.1

1. a $105°$ **2. b** $62°$ **3. c** $120°$

4. d $117°$, **e** $63°$ **5. f** $64°$ **6. g** $64°$, **h** $86°$

7. 144°   **8.** 89°

9. interior 108°, exterior 72°

10. 60°   **11.** 8   **12.** 40°

Page 34   Task M4.2

1. A and E

3. 81°

4. a $\binom{0}{3}$ **b** $\binom{-4}{4}$ **c** $\binom{4}{-1}$

5. 73°

6. a P and R   **b** 2

7. 135°

9. e.g.

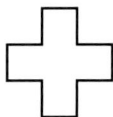

10. 30°

Page 37   Task E4.1

1. $2x - 180$ **2.** $y = 180 - x$

4. Andrea should have used brackets when subtracting expressions

5. $y = 3x - 30$

6. $180 - \dfrac{360}{n}$ or $\dfrac{180(n-2)}{n}$

7. $x - 45$

8. $180 - 4x$

Page 39   Task M4.4

1. SAS

3. a Alternate angles   **b** Vertically opposite angles

c AAS

4. a 16°   **b** 28°

c 136°   **d** Different reasons possible

6. Al has only proved that the triangles are similar

Page 41   Task M4.5

1. g $x = -1$

2. a x-axis   **b** $x = -1$ **c** $x = 2$

d $y = 1 \cdot 5$ **e** $y = -1$

4. f $y = x + 1$

Page 42   Task M4.6/M4.7

1. a $(3, 4)$ **b** $(7, 4)$ **c** $(4, 4)$

2. a 90° Clockwise centre $(0, 0)$

b 90° Anticlockwise centre $(3, -1)$

c 90° Clockwise centre $(1, -1)$

3. b $(1, 0)$ **c** $(9, 6)$

4. a 90° Anticlockwise centre $(4, 3)$

b $\binom{1}{-3}$

c 90° Clockwise centre $(6, 2)$

Page 43   Task M4.8

1. d $(6, -2)$

2. Scale factor 2, centre $(1, 5)$

3. Scale factor 3, centre $(-4, 5)$

4. Translation through $\binom{2}{-7}$

Page 44   Task M4.9

1. a Rotation 90° anticlockwise centre $(0, 0)$

b Reflection in x-axis

c Translation $\binom{3}{2}$

d Enlargement S.F. 3, centre $(0, 0)$

e Reflection in $x = -1$

2. g Translation through $\binom{-4}{-5}$

Page 45   Task M4.10

1. $\overrightarrow{AB} = \binom{3}{3}$, $\overrightarrow{CD} = \binom{-1}{-1}$, $\overrightarrow{EF} = \binom{2}{1}$, $\overrightarrow{GH} = \binom{-1}{3}$, $\overrightarrow{IJ} = \binom{4}{4}$

$\mathbf{a} = \binom{2}{3}$, $\mathbf{b} = \binom{-1}{-4}$, $\mathbf{c} = \binom{1}{2}$,

$\mathbf{d} = \binom{4}{-2}$, $\mathbf{e} = \binom{-3}{-5}$

3. $n = 6$

Page 46   Task E4.11

1. a $\binom{-9}{6}$ **b** $\binom{-20}{-4}$ **c** $\binom{-1}{6}$ **d** $\binom{-11}{3}$

e $\binom{15}{21}$ **f** $\binom{-15}{12}$ **g** $\binom{8}{12}$ **h** $\binom{-4}{1}$

2. a $\binom{5}{18}$ **b** $\binom{4}{8}$ **c** $\binom{-10}{0}$ **d** $\binom{19}{26}$

e $\binom{20}{-2}$ **f** $\binom{10}{1}$

3. c Q(1, −3) **d** R(3, 3) **e** $\begin{pmatrix} -2 \\ -6 \end{pmatrix}$ **f** $\begin{pmatrix} 3 \\ -4 \end{pmatrix}$

4. a −2 **b** 3 **c** 5 **d** 6

5. $\begin{pmatrix} -2 \\ 4 \end{pmatrix}$

Page 47   Task E4.2

1. a $\overrightarrow{BA}$ **b** $\overrightarrow{BF}$ **c** $\overrightarrow{AE}$ **d** $\overrightarrow{EC}$ **e** $\overrightarrow{BE}$

2.

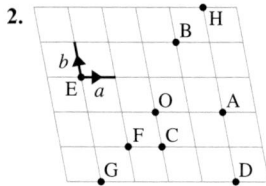

3. 2·8

4. 5·4

5. a 5 **b** 15 **c** 26 **d** 1·4

Page 47   Task E4.3

1. a $a + b$ **b** $-a - b$ **c** $a + b - c$ **d** $b - c$

2. a n **b** $-m$ **c** $n - m$ **d** $-m - n$

3. a (7, 3) **b** (6, 0) **c** $\begin{pmatrix} -1 \\ -3 \end{pmatrix}$

4. a $a + \frac{1}{2}b$ **b** $\frac{1}{2}a + b$ **c** $b - \frac{1}{2}a$ **d** $\frac{1}{2}b - \frac{1}{2}a$

5. $-q$

6. $\overrightarrow{MN}$ is a multiple of $\begin{pmatrix} 2 \\ 5 \end{pmatrix}$

Page 49   Task M5.1

1. b $3 \times 3 \times 3 \times 5$ **c** $3 \times 3 \times 5 \times 7$ **d** 45

2. −3, −4

3. $(-4)^2$ larger by 80

4. £32·39

5. £1258

6. 47

7. A greater by 7

8. 9

9. −4

Page 50   Task M5.2

1. $\frac{7}{9}$

2. a T **b** F **c** F **d** T **e** T **f** T

3. 121 000

4. $\frac{3}{20}$ m

5. Anna correct. 0·037 greater by 0·0324

6. £405

7. A gives 14

8. a $\frac{8}{11}$ **b** $\frac{29}{35}$ **c** −18 **d** $3\frac{2}{3}$

9. $2·17 \times 10^4$

10. $-\frac{43}{72}$ so need to add on $2\frac{43}{72}$

Page 51   Task M5.3

1. 1930 068

2. b and **d** are correct

3. Cheaper by £3 in Berlin

4. 4·62, 36·1, 0·579, 8·70

5. £80

6. Would need to use £61 of his saved money

7. Huan correct. It should be 4·47

8. £100

9. $\frac{5}{7}$

10. a 44% **b** 1 : 2 **c** 85·7%

Page 52   Task M5.4

1. a 33 **b** $58·4 \div 2 = 29·2$
   c $710 \div 2 = 355$ **d** $1560 \div 2 = 780$

2. a 4·5 **b** 34 **c** 5·6 **d** 0·48 **e** 4·3 **f** 6·3

3. 24

4. a $2·48 \to 124 \to 155$ **b** $5·8 \to 29 \to 72·5$
   c $9·7 \to 5·82 \to 19·4$

5. 0·6 m

7. C, D, A, B

Page 53   Task M5.5/M5.6

1. 15% **2.** 40% **3.** 30%

4. a 43% increase **b** 16% decrease **c** 95% increase

5. 20% **6.** 25·7% **7.** Joe by 1·6%

8. 52·1% **9.** 9% reduction

10. 3·2% increase **11.** 9·1%

Page 54   Task M5.7

1. 1·8 m   **2.** £120

3. £420   **4.** £340

5. 160   **6.** Hobbs by £1·40

7. 36   **8.** £230 000

Page 55   Task E5.1

1. £1272·50   **2.** No % change

3. 20% does not refer to £1200   **4.** £102

5. 1·8 cm^2 **6.** £3000

7. 216 cm^3

Page 56   Task M5.8

1. £8820   **2. a** £405   **b** £303·75

3. B   **4.** 7010 920

5. Tariq gains £6·02 more (accept £6·01)

6. 3 years

Page 56   Task E5.2

1. a 1·08   **b** £822·82

2. a 0·9   **b** £398·58

3. Supa Save by £92·93   **4.** During year 5

5. 10 years   **6.** 79 years old

7. £7089·19

Page 57   Task M5.9

1. a 6 : 1   **b** 12 : 5   **c** 1 : 3   **d** 8 : 21

2. £45 000   **3.** 3 : 8   **4.** 10 cm   **5.** 2·5 litres

6. 50   **7.** 12·8 m^2 **8.** 45 cm   **9.** 2 : 3

10. £$\dfrac{8n}{3}$

Page 58   Task M5.10

1. a 30   **b** 120   **c** 150

2. Yes

3. a 63   **b** 3

4. 22 m^2

5. b $y = 7x$ **c** $\dfrac{y}{4} = x$ **e** $y = \dfrac{1}{2}x$

6. a $k = \dfrac{36}{9} = 4$ **b** 28   **c** 20·25   **d** 46

7. F = kv

8. 22·5

9. $q = 8p$

10. a P = $\dfrac{2}{3}$F   **b** 18   **c** 24

Page 59   Task M5.11

1. a 15 hours   **b** e.g. the people work at the same rate

2. a 3·2   **b** 8

3. 9 hours

4. No

5. a 5   **b** $\dfrac{3}{4}$

6. b $y = \dfrac{4}{x}$,   **c** $xy = 6$,   **e** $4xy = 1$

7. a 2   **b** 0·5

8. $m = \dfrac{k}{p^2}$

9.

x	1	4	8	20
y	20	5	2·5	1

Page 60   Task M7.1

1. P : 4,   Q : -2,   R : 1,   S : $-\dfrac{1}{3}$

2. y : -7, -4, -1,   2,   5

3. a and **c**

4. a Stopped for some reason

   b 17·5 miles

   c 2:30 pm and 3:30 pm

7. b $y = 4x + 5$

8. Cerys walks 3 miles at a steady speed for 45 minutes. She then stops for 15 minutes. She then travels for another 30 minutes at a steady speed until she is 6 miles from her home then she returns home in a further $1\dfrac{1}{2}$ hours at a steady speed.

Page 62   Task M7.2

1. a $4x - 8$ **b** $12x + 6$ **c** $3m^2 - 2mn$

   d $n^2 + 4n$ **e** $3n^2 - 18n$ **f** $20m + 5m^2$

   g $3n^2 + 2n - mn$ **h** $2mn + 2m^2 - 8m$

2. a 21   **b** 26   **c** 41

3. a $x = y - 3$ **b** $x = 5v$ **c** $x = \dfrac{y + 2}{6}$

   d $x = \dfrac{y - 4}{2}$ **e** $x = \dfrac{y}{7}$

   f $x = 2(y - 9) = 2y - 18$ **g** $x = \dfrac{m + 3}{4}$

   h $x = 9(y + 2) = 9y + 18$

4. a 9   **b** 6   **c** 14   **d** 5

5. $3n + 7$

6. a $n^2 + 9n + 18$ **b** $y^2 - 6y + 8$ **c** $n^2 - 2n - 8$

    d $m^2 - 10m + 21$ **e** $p^2 + 4p + 4$ **f** $x^2 - 10x + 25$

7. 6 cm

Page 63 Task M7.3

1. a A3, B1, C4

2. a $a = 4, d = 3$ **b** $a = 6, d = 5$ **c** $a = 25, d = -3$

3. a $3n + 1$ **b** $5n + 1$ **c** $28 - 3n$

4. a $3n + 5$ **b** 95   **c** 25th term

5. a 2, 6, 12, 20   **b** 3, 9, 27, 81   **c** $-2, -2, 0, 4$

    d $\dfrac{1}{5}, \dfrac{1}{2}, \dfrac{7}{11}, \dfrac{5}{7}$

6. a $4n + 1$ **b** 217   **c** 35th term

7. a $5n - 3$ **b** $n = 8$

Page 64 Task M7.4

1. a 0   **b** 12   **c** 90

2. a 40, 53   **b** 34, 47   **c** 30, 42   **d** 66, 84

3. 144

4. 42   **b** 132

6. a, b, d are quadratic and **c** is arithimetic

7. $m + 2n, 2m + 3n, 3m + 5n, 5m + 8n$

Page 65 Task M7.5

1. a 3   **b** -4 **c** $\dfrac{1}{3}$ **d** 0·2

2. $-\dfrac{1}{2}$ **3.** 31 250   **4.** 5

5. a 36   **b** 324

7. a 2   **b** 384

8. 4

9. 1·05

10. a 3, 12, 48, 192   **b** Geometric with common ratio 4

Page 66 Task M7.6

1. Should also have divided c by m on line 2

2. a $x = \dfrac{y + w}{n}$ **b** $x = \dfrac{y - pq}{m}$ **c** $x = \dfrac{n - 4m}{c}$

    d $x = \dfrac{p - 2m}{q}$ **e** $x = \dfrac{7p + mq}{w}$ **f** $x = \dfrac{wy - ab}{3m}$

3. $mx - 3a = wy, \ mx = wy + 3a, \ x = \dfrac{wy + 3a}{m}$

4. a $x = \dfrac{3a - 5b}{w}$ **b** $x = \dfrac{fq + mn}{w}$ **c** $x = \dfrac{p - mw}{w}$

    d $x = \dfrac{5pw + ab}{m}$ **e** $x = \dfrac{y + 4m}{m}$ **f** $x = \dfrac{9p + fy}{f}$

5. $x = \dfrac{3mp + nw}{a}$

Page 66 Task E7.1

1. a $(a - b)^2$ **b** $a^2 - b$ **c** $\sqrt[3]{(6m - w)}$

2. a $\sqrt{(c - b)}$ **b** $\sqrt{\left(\dfrac{r - q}{p}\right)}$ **c** $\sqrt{\dfrac{b}{a}}$

    d $\sqrt{\dfrac{cb}{a}}$ **e** $(m + w)^2$ **f** $(zv - u)^2$

    g $\left(\dfrac{cd - b}{a}\right)^2$ **h** $\sqrt[3]{(a - b)}$ **i** $r^2 t^2 + p$

3. $\sqrt{(v^2 - 2as)}$

4. $\sqrt[3]{\dfrac{a}{4\pi}}$ **5.** $\dfrac{w - up}{4q^2}$ **6.** $\sqrt{\dfrac{A}{8}} + p$

7. a $\dfrac{m}{h}$ **b** $v(w + x)$ **c** $g(h - k)$

8. $\dfrac{c(h + a)}{b} - d$ or $\dfrac{c(h + a) - db}{b}$

9. $T^2 g + a$

10. a $h^2 m$ **b** $r^3 + z$ **c** $\sqrt[3]{\dfrac{2a}{m}}$

    d $\dfrac{c^3 + b}{a}$ **e** $\dfrac{p^2 - n}{m}$ **f** $\sqrt[3]{v - a}$

    g $\sqrt[3]{\dfrac{x + q}{p}}$ **h** $w^2 n + m$ **i** $\dfrac{27x^3 - p}{m}$

Page 67 Task E7.2

1. a $\dfrac{b}{a - c}$ **b** $\dfrac{3w}{p - 1}$

2. a $\dfrac{f}{m - c}$ **b** $\dfrac{w}{m - p}$ **c** $\dfrac{d - b}{a - c}$

    d $\dfrac{fb + ac}{a - b}$ **e** $\dfrac{my}{4 - m}$ **f** $\dfrac{mn - p}{q + m}$

3. $\dfrac{c}{a - b}$ **4.** $\dfrac{d}{a - e}$

5. $\dfrac{b}{ac - 1}$ **6.** $\dfrac{be}{k - e}$

7. a $\dfrac{a}{b + c}$ **b** $\dfrac{c}{1 - a}$ **c** $\dfrac{ab}{1 - b}$

8. $\dfrac{cu - dy}{ad - cb}$

Page 68   Task M7.7

1. a $<$ **b** $>$ **c** $>$ **d** $>$

2. a F   **b** F   **c** T   **d** F

   e F   **f** T   **g** T   **h** T

3. $4 \cdot 2$, $4\frac{1}{4}$, $4 \cdot 06$, $4\frac{1}{2}$

4. a $x > 6$ **b** $x \leqslant -3$ **c** $x < -1$

   d $2 \leqslant x \leqslant 6$ **e** $-2 < x \leqslant 3$ **f** $-4 < x < 0$

5.

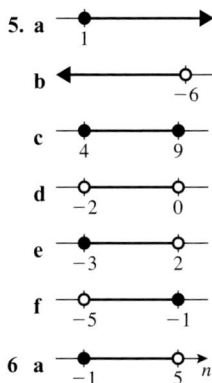

6 a

Page 69   Task M7.8

1. $x > 6$ **2.** $x < -5$ **3.** $x \leqslant 11$

4. $x < 6$ **5.** $x \geqslant 3$ **6.** $x > 18$

7. $x > 5$ **8.** $x \leqslant 5$ **9.** $x < 6$

10. $x \geqslant 6$ **11.** $x > 7$ **12.** $x \leqslant 24$

13. 3   **14.** 2

15. a 4, 5, 6, 7   **b** 1, 2, 3, 4, 5   **c** -1

   d $-3, -2, -1, 0, 1, 2, 3$

   e $-4, -3, -2$

   f $-5, -4, -3, -2, -1, 0$

16. a $-1 \leqslant x < 2$

   b $-3 < x < 3$

   c $5 \leqslant x \leqslant 11$

17. $5 \leqslant n \leqslant 8$

Page 70   Task M7.9

1. $x = 6$ or -3 **2.** $n = 0$ or -3

3. $a = 5$ or -1 **4.** $x = 1$ or 2

5. $a = -3$ or -1 **6.** $m = -5$ or -2

7. $y = 3$ or -4 **8.** $n = 2$ or -5

9. $x = 2$ or 6   **10.** $c = 3$ or -5

11. $m = 6$ or -4 **12.** $a = 3$ or 7

13. $n = 5$ or 0   **14.** $x = -7$ or 0

15. $y = 6$ or 0   **16.** $p = 2$ or -7

17. $n = 4$ or 8   **18.** $b = 3$ or 0

19. $x^2 + 6x - 91 = 0$, $x = 7 \, \text{cm}$

20. 3, 4

Page 71   Task M7.10

1. $-4 \cdot 8$ or $-0 \cdot 2$

2. a $-0 \cdot 6$ or $3 \cdot 6$ **b** $-0 \cdot 8$ or $3 \cdot 8$ **c** $2 \cdot 6$ or $0 \cdot 4$

3. a 0 or 2   **b** $2 \cdot 4$ or $-0 \cdot 4$ **c** $3 \cdot 2$ or $-1 \cdot 2$

Page 71   Task M7.11

1. a 6   **c** 2

2. a $(6, 0)(0, 4)$ **b** $(7, 0)(0, 3)$ **c** $(5, 0)(0, 8)$

   d $(6, 0)(0, -8)$

3. Yes

Page 72   Task M7.12

1. a $x = 4, y = 2$ **b** $x = 6, y = 6$ **c** $x = 2, \ y = 4$

2. d $x = 3, y = 2$

3. a $x = 1, y = 3$ **b** $x = 2, y = 5$ **c** $x = 3, y = 1$

Page 72   Task M7.13

1. $x = 1, y = 2$ **2.** $x = 3, y = 2$

3. $x = 2, y = 5$ **4.** $x = 5, y = 1$

5. $x = 3, y = 4$ **6.** $x = 0, y = 6$

7. $x = -1, y = 2$ **4.** $x = 4, y = -3$

9. $x = -2, y = -1$ **10.** $x = -5, y = 3$

Page 73   Task M7.14

1. $x = 3, y = 2$ **2.** $a = 1, b = 4$

3. $m = 2, n = 6$ **4.** $c = 5, d = 1$

5. $p = -1, q = 3$ **6.** $a = 4, b = -2$

7. $m = 3, n = \frac{1}{2}$ **8.** $x = -1, y = -2$

9. $x = -\frac{1}{2}, y = 3$

10 No answer possible. The lines never meet (i.e. are parallel).

Page 73   Task E7.4

1. Socks £3, pants £8
2. 7, 12
3. Battery £5, solar £6
4. Child £13, Adult £25
5. 45 cm
6. $m = 3$, $c = 5$
7. Hardback £5·95, Paperback £3·95
4. Anna 6, Charlie 18

Page 74   Task M7.15

1. a $(-3, 11)(-2, 6)(-1, 3)(0, 2)(1, 3)(2, 6)(3, 11)$
   b $(0, 2)$
2. a $(-5, 5)(-4, 0)(-3, -3)(-2, -4)(-1, -3)(0, 0)(1, 5)$ $(2, 12)$
   b i $(-2, -4)$ ii $(0, 0)$ iii $(-4, 0)(0, 0)$
3. a $(-5, -4)(-4, -8)(-3, -10)(-2, -10)(-1, -8)(0, -4)$ $(1, 2)(2, 10)$
   b i $(-2·5, -10·25)$ ii $(0, -4)$ iii $(0·7, 0), (-5·7, 0)$

Page 75   Task E7.5

1. a $(-3, -20)(-2, -3)(-1, 2)(0, 1)(1, 0)(2, 5)(3, 22)$
   b $(0, 1)$
   c $(0·8, -0·1), (-0·8, 2·1)$
2. a $(-2, -13)(-1, 0), (0, 3), (1, 2), (2, 3), (3, 12), (4, 35)$
   b $(0, 3)$
   c $(0, 3), (1·3, 1·8)$
3. e.g. quadratic curve not linear or equation suggests only one intercept with the x-axis
4. A → 5, B → 6, C → 2, E → 1, F → 3, G → 4
5. b $-1·5$

Page 76   Task M7.16

1. $5\,m^3/min$
2. 25 litres/week
3. b 40 km/h
4. a $2\,m/s^2$ b $1\,m/s^2$ c $0\,m/s^2$
5. $133\frac{1}{3}\,ml/s$
6. a $3·75\,m/s^2$ b $15\,m/s$ c $2·5\,m/s^2$

Page 77   Task M7.17

1. c, d, f
2. $y = 4x + c$

3. a $\frac{1}{3}$ b 4   c 1   d 4
4. a $(0, -4)$ b $(0, \frac{1}{3})$ c $(0, 4)$ d $(0, -1)$
5. Yes, each gradient $= 4$
6. Sam is correct
7. Not parallel, gradients are -2 and -1

Page 78   Task E7.6

1. $y = 6x + 5$ 2. $y = 2x + 3$ 3. $y = 3x + 10$
4. a $y = 4x + 1$ b $y = -3x + 2$
5. $y = -\frac{1}{2}x + 9$ 6. $y = 5x + 1$ 7. $y = 2x$
8. $y = 3x - 13$ 9. $y = -4x + 10$ 9. $y = -2x - 3$

Page 79   Task M8.1

1. 10
2. $\frac{7}{15}$
3. a $0·25, 0·325, 0·35, 0·325, 0·36, 0·39, 0·38, 0·37, 0·38, 0·38$
   c $0·38$
   d 228
4. a $\frac{1}{2}$ b $\frac{1}{2}$ c $\frac{1}{2}$
5. a $0·35$ b 6 times
6. $\frac{1}{6}$

Page 80   Task M8.2

1. No $\left(\text{Marie } \frac{4}{30}, \text{Don } \frac{6}{30}\right)$
2. $0·55$ or $\frac{11}{20}$
3. Natalie because everyone else got about the same answer.
4. a $\frac{1}{23}$ b $\frac{10}{23}$ c $\frac{2}{23}$
5. a $\frac{1}{2}$ b $\frac{64}{75}$
6. a $\frac{1}{8}$ b $\frac{3}{8}$
7. a $\frac{1}{2}$ b $\frac{11}{38}$
8. a $\frac{9}{20}$ b 15 times
9. a $120 : 80, 40 : 28, 52, 32, 8$ b $\frac{1}{2}$
10. $\frac{n - 6}{n}$

Page 82   Task M8.3

1. a {4, 6}   b {3, 5}   c 4
   d {3, 4, 5, 6}   e {4, 6, 7, 9}   f {7, 9}

2. a {2, 3, 5, 9, 10, 11, 12}   b 2
   c {2, 3, 5, 9, 10, 11}   d {5, 9}

3. a {ψ}   b {α, θ}   c 1
   d 3   e {β}   f {μ, γ, β, ψ}

4. a True   b False   c True

5. a {a, c, d, e, g, i} b 2   c {c, e, i, g}
   d {c, g}   e {a, b, c, d, f, g, h}
   f {a, b, c, d, f, g, h}
   g 3   h {e, i}   i {a, b, d, e, f, h, i}

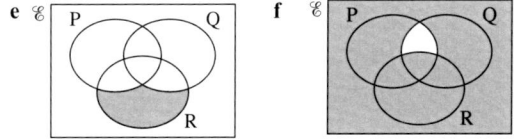

Page 83   Task M8.4

1. a   b

   c   d

   e   f

2. a P′ ∩ Q   b P ∩ Q′   c (P ∪ Q ∪ R)′

3.

4. a   b

   c   d

e   f

Page 84   Task M8.5

1. a $\frac{47}{80}$ b $\frac{59}{80}$ c $\frac{1}{16}$ d $\frac{21}{80}$

2. a $\frac{19}{40}$ b $\frac{1}{5}$

   c Probability of choosing a student who eats a school lunch but no school breakfast

   d $\frac{49}{200}$ e $\frac{4}{5}$

3. $\frac{7}{16}$

4. a $\frac{49}{83}$ b $\frac{11}{83}$ c $\frac{55}{83}$ d $\frac{16}{83}$

   e 12

5. $\frac{6}{25}$

Page 85   Task M8.6

1. $\frac{1}{4}$ 2. $\frac{1}{36}$

3. a $\frac{1}{16}$ b $\frac{1}{169}$

4. a $\frac{9}{121}$ b $\frac{4}{121}$

5. a 0·28   b 0·42   c 0·18

6. a $\frac{1}{4}$ b $\frac{9}{100}$ c $\frac{16}{100}$

7. $\frac{1}{1296}$

8. a $\frac{1}{7}$ b $\frac{3}{7}$

9. $\frac{n(10 - n)}{90}$

Page 86   Task M8.7

1. Part a only

2. a 0·7   b 0·2

3. $\frac{2}{5}$

4. $\frac{7}{12}$

5. a 0·15   b 0·4   c 8

6. Not exclusive

7. a 0·15   **b** 0·35

8. $\dfrac{13}{20}$

Page 87   Task M8.8

1. b $\dfrac{64}{121}$ **c** $\dfrac{9}{121}$ **d** $\dfrac{48}{121}$

2. b i 0·49   **ii** 0·42

3. b i $\dfrac{1}{64}$ **ii** $\dfrac{27}{64}$ **iii** $\dfrac{37}{64}$

4. a $\dfrac{5}{72}$ **b** $\dfrac{91}{216}$

5. a 0·008   **b** 0·384   **c** 0·488

Page 89   Task M8.9

1. b $\dfrac{5}{14}$ **c** $\dfrac{15}{28}$

2. b $\dfrac{7}{12}$ **c** $\dfrac{7}{18}$ **d** $\dfrac{11}{18}$

3. b i $\dfrac{11}{850}$ **ii** $\dfrac{997}{1700}$ **iii** $\dfrac{741}{1700}$

4. b $\dfrac{7}{44}$ **c** $\dfrac{37}{44}$ **d** $\dfrac{7}{22}$

5. a $\dfrac{(15-x)(14-x)}{210}$ **b** $\dfrac{15x-x^2}{105}$

Page 90   Task M8.10

1. a $\dfrac{9}{80}$ **b** $\dfrac{5}{8}$ **c** $\dfrac{1}{5}$ **d** $\dfrac{59}{80}$

2. b 0·375

3. a $\dfrac{2}{5}$ **b** $\dfrac{1}{5}$

4. b 63%

5. a $\dfrac{52}{147}$ **b** $\dfrac{1}{7}$ **c** $\dfrac{11}{49}$ **d** $\dfrac{1}{21}$

   e Probability of having an indian meal but no chinese meal

6. $\dfrac{3}{22}$

Page 92   Task M9.1

1. 23 minutes   **2.** 2068 kg   **3.** 3750 g

4. 21 km/h   **5.** Motorbike faster (72 km/h)

6. 15:04

7. A 50·4 km/h, B 54 km/h

8. 8 miles

9. No. 2 litres is reasonable in one day.

10. 41 mph

Page 93   Task M9.2

1. a 42 cm²   **b** 78·5 cm²   **c** 75 cm²

2. 3927·0 cm²

3. 6 cm

4. 79·2 m

5. £17 898

6. Both shaded areas are equal

Page 94   Task M9.3

1. 754·0 cm³   **2.** 3250 cm³   **3.** 3960 cm³

4. 8 cm   **5.** 10   **6.** £13·26

7. 32.0 cm

Page 95   Task M9.4

1. a 23·35 cm   **b** 23·45 cm

2. a 3·75 m   **b** 3·85 m

3. 62·5 kg

4. $47·15 \leqslant l < 47·25$, $82·5 \leqslant m < 83·5$, $7·25 \leqslant V < 7·35$, $6·865 \leqslant r < 6·875$, $465 \leqslant A < 475$

5. a 3·45 cm   **b** 4·75 cm

6. $503·35 \leqslant c < 503·45$

7. $23·455 \leqslant \text{time} < 23·465$

8. Incorrect, 8450 would be rounded up to 8500

9. $0·00175 \leqslant \text{width} < 0·00185$

10. £950 000

Page 96   Task M9.5

1. a 10·61   **b** 29·32   **c** 7·68

2. a $4·236 \leqslant x < 4·237$

   b $715·187 \leqslant x < 715·188$

   c $26·548 \leqslant x < 26·549$

3. 193·375 cm³

4. $8·2 \leqslant m < 8·3$

5. 12·65 cm² (L), 13·20 cm² (H)

6. Amelia (64 kg), John (63·5 kg), Amelia sinks to the ground

7. $4700 \leqslant n < 4800$

8. a $350 \leqslant d < 370$ **b** $42·5 \leqslant m < 47·5$ **c** $6·3 \leqslant x < 6·5$

9. 7

Page 97   Task M9.6

1. 9 g/cm^3
2. 480 g
3. 30 cm^3
4. 5.25 Pa
5. 3.2 m^2
6. Steel by 0.5 cm^3
7. 35.52 kg
8. 20.1 N
9. 62.38 kg
10. 30 cm by 20 cm face

Page 98   Task E9.1

1. $\dfrac{18x}{5} \text{ km/h}$
2. $\dfrac{q}{\left(\dfrac{x}{100}\right)^2} = \dfrac{10\,000q}{x^2}$
3. 74.5 km/h
4. $63(x^2 - 3y)$
5. $\dfrac{5}{18}(5x - 3y)\text{m/s}$
6. 55
7. $\dfrac{mxy}{200}$
8. $\dfrac{1000m + n}{\dfrac{1000m}{x} + \dfrac{n}{y}} = \dfrac{xy(1000m + n)}{1000my + nx}$

Page 99   Task M9.7

1. 4.8 cm
2. 17.0 cm
3. $8 + \dfrac{2}{3}\pi$
4. $4\pi + 36$
5. $\dfrac{10\pi}{3} + 14$
6. Pete is correct
7. $43°$
8. 20.6 cm
9. 36.8 cm
10. 33.2 cm

Page 101   Task M9.8

1. 47.1 cm^2
2. 51.5 cm^2
3. 34.7 cm^2
5. $40 - \dfrac{32\pi}{9} \text{ cm}^2$
6. $17.9°$
7. 7.1 cm^2
9. 44.6 cm^2

Page 102   Task M9.9

1. a 800 cm^3 b 0.0008 m^3
2. a 7.2 m^3
3. False
4. 5 m
5. a $4\,000\,000$ b $2\,900\,000$ c $80\,000$
   d $74\,800$ e 6 f 6000
   g 600 h 5160 i 3.8
6. 7.5
7. 70 cm

Page 104   Task M9.10

1. a 1940 cm^3 b 28 m^3 c 65.4 m^3
2. The cone
3. a $1488\pi \text{ cm}^3$ b $2511\pi \text{ cm}^3$
4. 2 min 7 secs   5. 2160π grams
6. 8.2 cm 7. 2.67 cm

Page 105   Task M9.11

1. a 239 cm^2 b 1020 cm^2 c 1010 cm^2
2. $90\pi \text{ cm}^2$ 4. 754 cm^2 5. 2.39 cm
6. a 6 cm b $108\pi \text{ cm}^2$
7. 1190 cm^2 8. 12.4 cm 9. 2810 cm^2

Page 106   Task M9.12

1. b 9 cm
2. 10.5 cm
3. $y = 12.5 \text{ m}, z = 11 \text{ cm}$
4. b 25 cm
5. b 3 cm
6. a 10 cm b 5 cm

Page 108   Task E9.2

1. 2.5 cm 2. 9 cm 3. 18 cm
4. $AB = 8 \text{ cm}, AE = 4 \text{ cm}$ 5. b 5 cm
6. b 10 cm 7. $x = 5 \text{ cm}, y = 7 \text{ cm}$
8. 5 9. $x = 8 \text{ cm}, y = 15 \text{ cm}$
10. 216 cm^2

Page 109   Task M10.1

1. a

	Paper	Bottles	Cans	Total
Boys	86	73	89	248
Girls	220	58	74	352
Total	306	131	163	600

   b 220 c 27.2%
2. 378
3. Angles $112°, 68°, 32°, 132°, 16°$
4. b £24 500   c £6825
5. £4.78

6. a

Stem	Leaf
10	7 7 9 9 9
11	5 6 6 7 7 9
12	2 2 2 3 4 4
13	3

$12|3 = 12 \cdot 3$

b 0·5 cm (median = 11·7 cm)

7. 2, 8, 12, 14, 14

Page 111   Task M10.2

1. Use e.g. Grafton median = 23, range = 6 and Colby median = 28, range = 16

2. £415

3. Not enough info for Arjun to make the statement

4.

	Car	Walk	Bike	Train	Total
Birmingham	314	117	31	69	531
Nottingham	216	175	41	37	469
Total	530	292	72	106	1000

a $\dfrac{31}{531}$ **c** 10·6%

5. Angles 30°, 50°, 80°, 120°, 80°

6. Frequencies: 4, 6, 7, 2

7. a 66   **b** $n^2 + 2$

Page 112   Task M10.3

1. a 4   **b** 3   **c** 2·78

2. a A 3·4   B 4·3   **b** Area B

3. Use e.g. 11B: median = 7, range = 6, 11A: median = 5, range = 9

4. a 670   **b** 3·35

5. Correct

6. 12

Page 114   Task M10.4

1. b 240

c 250, 150, 145, 55, 105, 210, 340, 475

d Sales fell after March and then rose quickly from September onwards.
More card sales for Christmas, Valentines Day and Easter

2. b 105, 110, 105, 120, 125, 135, 135, 125, 160, 165, 135, 140, 125

d Sales rose 2003 to 2011 (apart from 2008) and then fell a little.

3. b 130, 125, 125, 127, 127, 128, 130, 131, 132, 131, 130, 132, 137, 135, 137

d Number of visitors rose gradually.

Page 115   Task M10.5

2. c (200, 13)   **d** Positive

e 7 cm   **f** 193 − 195 cm

4. b Negative   **d** 72

e Not within the range of given data and the average weekly score would be impossible

Page 116   Task M10.6

1. a Yes

b Not representative. The sample may not include poeple who need a car to go to the supermarket.

c Yes

d No. The sample only takes people who use the gym.

e Yes

f Yes

2. a Random sample of people under 18 years old. Include people aged (say) 11 to 17.

b Random sample of people in Scotland.

c Random sample of pupils in school.

d Random sample of people in the city.

Page 117   Task M10.7

1. 31 males, 19 females

2. Ski 22, AUST 8/9, EU cities 25, Carib 14/15

4. Lab. 706, Con. 539, Lib. Dem. 428, Green 310, Other 17

5. 90

6. b F 14, G 7, S 4

Page 118   Task M11.1

1. a 76°   **b** 93°

2. a 5·4 cm   **b** 10·7 cm

3. 7·6 km

5. b 9·8 m   **c** 157 m²

6. 2·7 cm

7. 6·2 cm

Page 119   Task M11.2

1. 17

2. b 7 km   **c** 232°

3. 52 cm²

4. a 208°   **b** 061°   **c** 304°

5. a D(3, 1)   **b** (−1, 1)   **c** 12

6. two sides will not match (2 and 4)

7. 050°

8. b (1·5, 2·5)   **c** N(6, 1)

Page 121   Task M11.3

5. d QS = 5 cm

Page 122   Task E11.1

5. AB = 10·4 cm   **6.** x = 5·9 cm   **8.** AD = 7·6 cm

9. a 3·7 cm/3·8 cm

Page 125   Task M11.6

4. B (32 cm²) greater by 4 cm² than A

5. a 4   **b** 6

6. e.g.

front   side   plan
elevation   elevation

Page 127   Task M11.7

1. 7·81 cm   **2.** 5 cm

3. a 4·47 cm   **b** 10·30 cm   **c** 18·57 cm

4. 9·22 cm   **5.** 13·23 cm

6. 11·40 cm   **7.** 48 m

8. 5·10 m   **9** 180·28 km

10. 270 cm²

Page 128   Task E11.2

1. 5·39 cm   **2.** 5 cm   **3.** 19·42 cm

4. a 8·49 cm   **b** 11·31 cm

5. 49·61 cm²   **6.** 42·60 cm

7. 424·12 cm²   **8.** 9852·03 cm²

9. 17·49 m   **10.** 1·41 cm

Page 130   Task M11.8

1. No   **2.** Yes   **3.** No   **4.** No

5. Yes   **6.** Yes   **7.** Yes   **8.** No

Page 131   Task M11.9

1. 10·5 cm   **2.** 17·3 cm   **3.** 8·02 cm

4. 22·2 cm   **5.** 7·58 cm   **6.** 11·3 cm

7. 14·5 cm   **8.** 27·6 cm

9. a 9·20 cm   **b** 10·9 cm

10. 15·8 cm   **11.** 20·7 cm   **12.** 7·66 cm

Page 132   Task M11.10

1. 62·2°   **2.** 25·3°   **3.** 79·9°

4. 28·9°   **5.** 44·3°   **6.** 14·0°

7. 10·3°   **8.** 65·4°   **9.** 64·4°

10. a 10 cm   **b** 89·3°

11. 25·9°

Page 133   Task E11.3

1. a 45°   **b** $\sqrt{2}$ cm   **c** $\dfrac{1}{\sqrt{2}}$ **d** 1

2. 6 cm

3. a 60°   **b** 1 cm   **c** $\sqrt{3}$ cm

   d i $\dfrac{1}{2}$ **ii** $\sqrt{3}$ **iii** $\dfrac{\sqrt{3}}{2}$

4. 5 cm

5. $7\sqrt{3}$ cm

6. $\sin 30° = \dfrac{1}{2}$, $\cos 30° = \dfrac{\sqrt{3}}{2}$, $\tan 30 = \dfrac{1}{\sqrt{3}}$

7. 6 cm

8. a $(24 + 8\sqrt{3})$ cm   **b** $(30 + 10\sqrt{3})$ cm

10. 60°

Page 135   Task M11.11

1. 49·5°   **2.** 16·5 cm   **3.** 10·8 cm

4. 21·2 cm   **5.** 51·4°   **6.** 3·14 cm

7. 1·88 m   **8.** 7·34 cm   **9.** 5·91 m

10. 11·0 cm

Page 136   Task E11.4 _____

1. $29 \cdot 4 \, \text{cm}^2$ 2. $13 \cdot 8 \, \text{cm}^2$

3. $10 \cdot 8 \, \text{km}$ 4. $17 \cdot 9°$

5. $(21 + 7\sqrt{3}) \, \text{cm}$

6. **a** $28 \, \text{cm}$ **b** $1344 \, \text{cm}^2$

7. $7 \cdot 48 \, \text{cm}$ 8. $32 \cdot 5 \, \text{cm}$